THE VOODOO PROJECT

George Overton

An Inner Workings Publication

The circuits presented in this book are for experimental purposes only. Some designs may be covered under active patents and may not be commercially developed without permission from the patent holder. Some of the circuits may also fall under government regulation for radio emission. It is the responsibility of the user to ensure adherence to these regulations.

Some studies have indicated that devices which emit radio frequencies may be linked to certain health problems, including disruption of pacemakers. Although there has been no indication that metal detectors are among these devices, health safety is the responsibility of the user.

Neither the author nor the publisher of this book shall be held responsible, whether legal, financial, or any other way, for any damages incurred by the use of the contents of this book.

This book was put together using LibreOffice Writer, with equations written using LaTeX via the TexMaths extension. Body text in Times New Roman 10pt, URW Chancery L (various sizes) for titles, and Flipside BRK for the book title.

Line art was drawn with LibreOffice Draw, and images manipulated with GIMP.

Circuit diagrams and PCB layout were generated with Easy-PC from Number One Systems, and simulations were done with LTspice.

Embedded software was created with Mikroelektronika's mikroC PRO compiler, and PIC processors were programmed with MPLAB PICkit4 via MPLAB X IPE.

3D printed parts were initially created with FreeCAD, then stl files were sliced in Repetier-Host using CuraEngine, and finally printed with a GEEETech i3 Pro B.

Both Linux Mint and Microsoft Windows operating systems were used during the development of the Voodoo Project, and to create the contents of this book.

Table of Contents

"Science ... because figuring things out is better than making things up."

--- Anon

Introduction

"If there are no stupid questions, then what kind of questions do stupid people ask? Do they get smart just in time to ask a question? "

--- Scott Adams

About this Book

This book is a written record of the author's personal mission to design and develop a working pulse induction (PI) metal detector that is capable of rejecting iron.

Here are some typical quotes concerning pulse induction:

"Pulse Induction metal detectors are specialized instruments. They are generally not suitable for coin hunting urban areas because they do not have the ability to identify or reject ferrous (iron) trash. "

"Due to the nature of the signals caused by different objects, it is extremely difficult, or put another way, almost impossible to build a good discriminating PI. Since the time it takes for a target signal to decay can vary because of the size, shape, and chemical makeup of the object, then any type of later sampling will not produce a reliable form of discrimination." - Reg Sniff

"The various models of PIs are so different from one another, it is difficult to generalize about how threshold should be used, or what kind of discrimination they offer. It can be said that none of the PIs from major manufacturers offers the sort of discrimination features which are found on many VLFs, because PIs do not readily distinguish between ferrous and non-ferrous metals. It can be said that the ground balancing methods used in PIs, which make them so effective in 'hot ground', tend to reduce response to metallic iron and also to reduce response to non-ferrous metal targets in a certain size range. The poor discrimination characteristics of PIs is a primary reason why no general purpose

metal detectors use this principle. This essay reflects the industry state of the art as of April 2009." - Dave Johnson, Chief Engineer, First Texas and Fisher Research Labs.

Anyway, you no doubt get the idea, but that never stopped anyone from trying to prove them wrong. In the next chapter we will explore the state-of-the-art regarding PI discrimination, and discuss their limitations. There are also several patents available, which outline methods for providing ferrous/non-ferrous discrimination, but none of these have ever made it into production. Often there are rumours of an "ultimate" detector that combines the features of both PI and VLF, but these always seem to fade away over time. Although there are numerous reasons why these detectors never saw the light of day, not all of them were technical.

The infamous *YouTube* is a good place to watch videos of various home-made (and commercial) PI detectors being tested, with many of them demonstrating how great they are at rejecting ferrous (iron) targets. To the uninitiated, it may seem like the problem has already been solved, but closer examination reveals the limitations of the testing and (in certain cases) the use of some "smoke and mirrors". It's usually not what you see in the video that's important, but what you don't see.

Beyond the patents and practical experiments there are even more theoretical propositions that never made it off the drawing board. Unfortunately the journey from concept to a real physical PI detector, that can reliably reject iron, is one that is littered with many failures.

Acknowledgements

Carl Moreland and I wrote editions 1 and 2 of "*Inside the* METAL DETECTOR" (often referred to as ITMD) as a collaborative effort. As a result we both learned a lot from the exercise. Previously we dedicated the book to the many contributors to the Geotech forum (www.geotech1.com), and also to Eric Foster who continues to help people understand pulse induction techniques. Hence I would like to do the same here. Also, I would like to thank my long suffering wife, who thinks I spend far too much time "playing around with bits of wire".

Some Assumptions

The fact that you've purchased this book *hopefully* means that you understand at least something about the inner workings of both PI and VLF detectors. If not, then [hint] ITMD would be a good place to start, otherwise you will most likely struggle to understand some of the concepts in this book.

If you are intending to experiment with the concepts discussed in this book, then please be aware that you will need to have access to some reasonable test equipment. Without a 2-channel oscilloscope, a multimeter, and an LCR meter, your attempts will be doomed to failure. Plus you will need knowledge of embedded C programming, and some considerable patience.

Please do not even attempt this project without having built a standard (PI and/or VLF) metal detector before. There is no substitute for knowing what you're doing.

Just to make certain that you fully comprehend the importance of what's been written above:

1. If you are a beginner in electronics, then forget it. You won't understand what's being discussed in this book.

2. If you've never built a metal detector before, then go away and build one, or preferably more than one. At a minimum, a VLF and a PI.

3. If you don't understand what VLF and PI mean, then read "*Inside the* METAL DETECTOR" first.

I apologize for being so blunt, as my intent is not to deter you from reading this book. In fact I urge you to do so. However, if you don't know what you're doing now as regards electronics, and metal detectors in particular, then you'll probably be wasting your money.

Having said that, enjoy the rest of this book, and thanks for reading.

Chapter 1 _____ *Current State-of-the-Art*

"By the pricking of my thumbs, something wicked this way comes."

--- 2nd Witch – Macbeth, Act 4, Scene 1

Pulse Induction Technology

PI detectors operate in the time domain by energizing a coil with a DC current. This produces a static magnetic field which is switched off very quickly, causing the magnetic field to collapse. The higher the current, and the faster it is switched off, the higher the transient magnetic field. When the battery is connected across the coil the current rises in an exponential manner until it reaches a maximum value that is dependent on the coil's DC resistance, any series resistor, the switch and the battery. While the switch is closed the coil generates a magnetic field. When the switch is opened the current will drop to zero, and (as a consequence of Ohm's Law) the flyback voltage across the coil becomes extremely high. In fact it reaches several hundred volts. One interesting thing to point out is that the coil current does not reverse direction (as some people erroneously think) but continues to flow in the same direction. In essence, it increases during switch-on, and decreases during switch-off. When the switch turns off the coil changes from a load into a generator, and the impedance of the coil becomes negative. The result is that the flyback voltage reverses polarity. Hence if the applied voltage during switch-on is positive, the flyback voltage will have a negative polarity, and vice versa. The coil's magnetic field also collapses, and any metallic objects close by will have eddy currents generated within them due to flux linking.

Flux Linking versus Flux Cutting

Flux cutting occurs when the geometry (the relationship between a magnet and a loop of wire) changes. For example, if you wave a magnet over a coil, or move the coil past a magnet. *Flux linking* occurs when the geometry remains stationary, but the magnetic field itself changes. For example, flux linking occurs in a transformer. Of course, you could argue that a metal detector coil *does* move in relation to a metal target, but the primary cause of eddy currents is flux linkage. Flux cutting is a secondary effect. Remember that some detectors are non-motion.

Unlike the smooth exponential curve of the coil current during switch-on, the decay curve of the current during switch-off exhibits ringing (oscillations) due to the coil's distributed inductance and capacitance, and the overall slope of the ringing follows an exponential decay defined by the coil resistance. It is necessary to place a damping resistor across the coil to achieve what is called critical damping. The intention is to reduce the amplitude to zero in the fastest possible time with zero oscillations. Although an underdamped coil can reach zero amplitude more quickly than a critically damped one, it will oscillate around zero. Whereas an overdamped coil will approach zero amplitude more slowly. This is shown in Fig. 1-1.

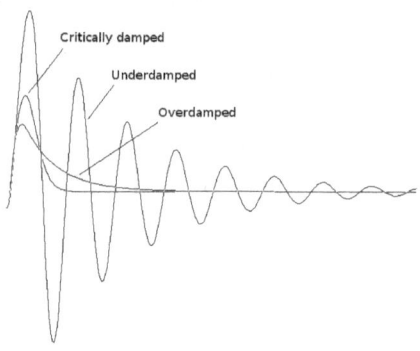

Fig. 1-1: Critical Damping versus Underdamping and Overdamping

Although it is possible to calculate the value of the damping resistor, it is best found by trial and error.

The eddy currents, which are generated in any nearby magnetic targets, generate their own magnetic fields, and these have an effect on the decay curve. The changes in the slope of the curve are miniscule, in the region of milli or even micro-volts. The trick is to sample the decay curve before the target eddy currents have died away. Again, ITMD goes into much greater detail than is necessary here.

Using this method of detection, there is some information regarding the nature of the target, as the tau (τ) affects the slope of the decay curve. The target tau is also known as the target decay constant.

PI detectors are superior to VLF detectors regarding depth in areas having a higher ground mineralization. However PI detectors are more susceptible to external noise because the preamp is a broadband amplifier. For the purposes of this project, a PI detector is defined as one that conforms to the previous description. That is, a detector that has discrete on and off times, with one or more samples being taken during the off period.

VLF Technology

Whereas the PI (according to our previous definition) uses a discrete-time transmitter, VLF detectors use continuous-time. There are no periods of zero current in a VLF detector. As a consequence, VLF detectors carry out their sampling while the transmitter is still active. The metal detector world is riddled with 3-letter acronyms (TLAs), and this causes a huge amount of confusion. In particular, some of Minelab's detectors tend to blur the lines between PIs and VLFs, and the confusion is not helped by some of the marketing literature.

The Difference

The differences between VLF and PI detectors are due to fundamentally different technologies. VLF detectors use balanced coils, which are essentially a loosely coupled transformer system. Any metal target will cause an imbalance in this [otherwise-balanced] system and produce an audio response, whereas a PI detector relies solely on eddy current detection in the target. This seems to be the case because a PI uses much greater amplification than a VLF. In fact so much so that it is necessary to provide a method of eliminating false responses due to the coil being moved within the Earth's magnetic field.

If you take a VLF detector that has been perfectly ground balanced and then use it in neutral soil, there will be no response from the GB channel. However, the DISC channel is still being affected. Motion detectors reduce this effect by only detecting fast changes due to metal targets and suppressing any slow changes due to the ground matrix, or relative movement between the coil and the ground surface. For a PI, the target signal is unaffected by the same neutral soil. The soil appears to present less losses to a PI due to the different detection mechanism.

VLF Definition

If you want to be pedantic, then you could point out that not all VLF detectors use a balanced coil system, commonly referred to as induction balance (IB). In that case I guess we should really be comparing PI detectors (using either mono or IB coils) with VLF detectors (with IB). Confused? You're not the only one.

PI – What's available now?

If you want to purchase a PI detector and expect to get some form of target discrimination, then what's available? To start with, any PI detector where you can vary the main sample delay can provide variable rejection of low conductivity targets. This is accomplished by moving the main sample pulse to a position where the target's eddy currents have already died away.

Going back to the 1980s, there was a PI detector known as the Minipulse that was designed by Eric Foster for Pulse Technology Ltd., from Abingdon, Oxon, England. This detector had a very low TX pulse rate of 86pps, and the control box had only two rotary knobs. The first was the on-off /audio control, which was used to turn the unit on-off and to set the audio threshold. The second knob was marked as REJECT, with foil and ring pull (pull tab) positions marked. Unlike a VLF detector, where iron is the first target to get rejected, a PI detector cannot reject iron. This is due to the fact that PIs rely on detecting eddy currents, whereas VLFs are more dependent on the effects of absorption and redistribution to unbalance the coil. So the main difference between the two types of detectors is that the VLF can tell the difference between ferrous and non-ferrous targets, whereas the PI can only discriminate based on conductivity.

However not all PI detectors have the facility to adjust the main sample delay. For example, C.SCOPE's latest PI offering (the CS4Pi) is designed as a switch-on-and-go unit, and is intended to be an all-metal detector. As a consequence, it has no rejection capability.

By contrast, White's Electronics PulseScan TDI has many more adjustments. Its Pulse Delay control can be used to adjust the main sample delay from a minimum of 10µs up to 25µs. When set to 10µs this gives the detector the highest sensitivity to all targets regardless of conductivity. Raising the delay towards 25µs maximizes the signal to high conductive metals while suppressing the signal from low conductive metals. There is also a Ground Balance control for balancing out noise from ground mineralization. In situations where the ground mineralization is low, it is possible to hijack the ground balance control to eliminate a selected target based on its conductivity. For example, small iron nails. Whereas the Pulse Delay control will ignore targets with a conductivity below a certain setting, the Ground Balance control is more like a notch feature. Any target below the notch will give a high tone, and any target above the notch will give a low tone. All other targets that fall within the notch (often referred to as a target hole) will not be heard. This "trick" only works for a selected target conductivity, and for targets of a very similar size and shape. For example, if you adjust the Ground Balance control to reject a small rusty nail, then other iron targets of differing sizes and shapes will still be heard. The use of either the Pulse Delay and/or the Ground Balance control are examples of target discrimination by conductivity. As you can see, it is not possible to eliminate ferrous (iron) targets with the same effectiveness as a VLF detector.

Some people have theorized that it may be possible to take several samples along the decay curve, and potentially provide ferrous/non-ferrous discrimination in that way. Plotting the decay curves for several different targets shows that this approach does not work, and can only provide a guess as to the target type, again based on conductivity.

One solution to the problem is that taken by a Bulgarian company with their Grizzly Twins detector. This machine combines a PI and VLF detector in the same box. The controls for the VLF detector are on the left-hand side of the panel, and the PI detector controls are on the right-hand side. There is a button in the middle of the panel that switches between the two modes. Initially this sounds like a neat solution, but it appears that it is necessary to swap over coils in order to use each detector. The general idea is that you first search in VLF mode to clear away all the near-surface objects, and then swap over the coils and use the PI mode to search for anything deeper.

There are a few commercial offerings that provide a PI detector with a built-in magnetometer. These detectors use large 1m or 2m frames to search for large deeply buried objects while ignoring iron and steel.

As far as patents are concerned, there are the half-sine and truncated half-sine methods. The half-sine method is patented by Barringer Research, and truncated half-sine is from White's Electronics. The general idea is that these two methods provide intervals where the transmitter is in an off state. In addition, the truncated half-sine method includes a faster turn-off on portions of the transmitted signal to improve the pulsed response when compared to half-sine. Although the half-sine method has been used extensively in aerial mapping of ore deposits neither of these methods have yet made it into a commercial hobby product, so their effectiveness at providing ferrous/non-ferrous discrimination is generally unknown.

Sometimes you will come across metal detector kits that boldly claim an ability to reject iron. If the kit is a true PI circuit, then it always turns out that the so-called iron-rejection is based on target conductivity, and any *YouTube* videos are very selective in the choice of targets being rejected. If the detector being shown is using a mono coil, then you can be doubly certain of the fact. You also need to be wary of some detector kits with the word "pulse" in the name, and the assumption by some people that this means "pulse induction". One particular *YouTube* video shows such a detector kit with a visual display indicator (VDI) that is clearly not possible with a standard PI. In reality, the word "pulse" refers to the fact that it uses a transmitter with a continuous square wave output, and is really a VLF design.

Project Goal

As far as this project is concerned, we obviously do not want to use target conductivity in a vain attempt to provide ferrous/non-ferrous discrimination, and any method that is covered by a patent is also out of the question. To be clear – the goal of this project is to design and build a PI detector aimed at coin shooting, and for use on both the beach and inland. At this time we are not attempting to create a machine capable of looking for tiny gold nuggets in the ghastly ground conditions that Australians refer to as "soil". Hence ground balance considerations are initially not being considered, as PI detectors are quite capable of ignoring low and even medium levels of mineralization in their own right, especially if not specifically searching for very low conductivity items such as gold nuggets.

So, how to proceed? That is the subject of the next chapter…

Chapter 2 _Let the Experiments Begin_

"Welcome to the nightmare. I've been expecting you."

--- Human Zoo (A Letter From Death Row) – Bret Michaels

Having constructed Carl Moreland's Hammerhead PI design, and also designed and built Raptor (a VLF design from ITMD) I wanted to experiment with the idea of combining the two in such a way as to provide ferrous/non-ferrous discrimination.

The basic idea was to drive the outer TX loop with a PI transmit circuit and to receive from the same loop. In this case we would have a PI detector with a mono coil. The inner RX loop could then be made to resonant by adding a parallel capacitor.

Raptor was designed to use Tesoro 5-pin coils with a TX inductance of 6mH, and an RX inductance of 6.5mH. With a TX inductance of 6mH this was obviously far too high for a PI detector, but I did have a 4-pin Troy 9" diameter concentric coil available. The TX inductance of this coil was 1.12mH with an RX value of 6.96mH. Although 1.12mH was higher than I initially desired it would be sufficient for some exploratory tests. At least it would allow the use of a commercial coil and alleviate the potential problems associated with constructing my own coil.

The outer TX loop was connected to a Hammerhead circuit with a parallel damping resistor to achieve critical damping, and a 33nF capacitor was placed in parallel with the RX loop. This caused the signal across the RX loop to resonate at 10.5kHz. Prior to the experiment, there was some concern that the RX loop would look like a highly conductive target and could result in some associated ringing in the TX loop. Examining the output of the Hammerhead preamp showed that the Troy coil was extremely well balanced, and none of the RX ringing was visible. Using an oscilloscope, the signal at the RX loop was observed to be in the form of a decaying sine wave, and the signal amplitude moved in different directions depending on whether the target was non-ferrous or ferrous.

As can be seen from Fig. 2-1, the maximum RX amplitude on the first half-cycle of the sine wave was approximately 10mV peak after the MOSFET switched off.

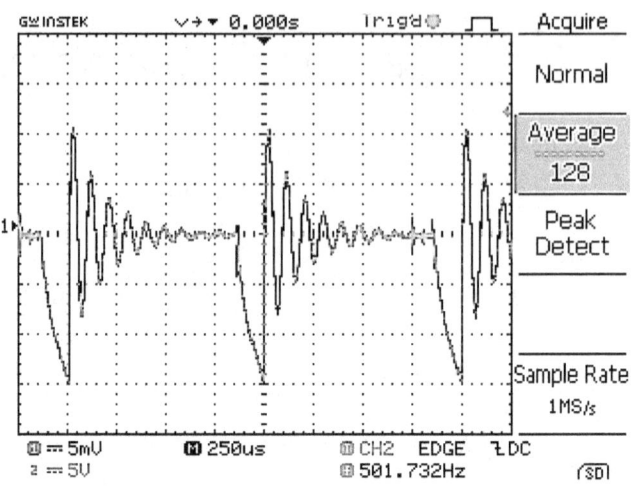

Fig. 2-1: Decaying Sine wave at RX Inner Loop

The next step in the experiment was to feed the RX signal from the inner loop into a second preamp.

At that time I had been experimenting with PIC micro-controllers, and had purchased a Mikroelektronika EasyPIC5 development system and their mikroBasic compiler. The PIC16F877A that came with the development system was programmed to generate the necessary pulses required for synchronous demodulation of the signals from the two preamps. From here on, these two signal paths will be referred to as the PI and DISC channels throughout this book.

In the same way as the PI channel used one main sample per TX pulse period, the DISC channel used one sample that encompassed the second half-cycle of the decaying sine wave. This was equivalent to the quadrature demodulation sample in a VLF detector. Whereas the PI demod output always had the same polarity regardless of target type, the DISC demod changed polarity depending on whether the target was ferrous or non-ferrous. Clearly (at least on the bench) it was possible to eliminate iron with this approach.

As you can see, the plan was to replace the VLF detector's GB channel with a PI channel. The main reason being that a VLF detector must be ground balanced, even when the ground is non-mineralized. When the coil of a VLF is lowered to the soil, signal

The Voodoo Project

absorption occurs and this causes a change in the amplitude of the received signal, but without any phase shift. Since the GB channel is sampling the RX signal at (or close to) the zero-crossings, it effectively ignores this ground response. On the other hand, a PI detector naturally ignores neutral (or even light to medium mineralized) ground, as any eddy currents have effectively died away before sampling occurs.

Having proved the basic concept on the bench, it didn't take long to conclude that a dedicated board would be required for any serious testing to commence. Hence the next step was to construct a Heath Robinson prototype on Veroboard (stripboard) with the PIC micro, a 2-line 16-character LCD, and the necessary electronics. An in-circuit system programming (ICSP) interface was also included.

Heath Robinson

William Heath Robinson (1872 – 1944) was an English cartoonist, illustrator and artist. He was best known for his drawings of whimsically elaborate machines to achieve simple objectives. In the UK, the term "Heath Robinson" is used to describe any unnecessarily complex and implausible contrivance, or more often an ingenious contraption constructed with whatever comes to hand.

In the USA, the term "Rube Goldberg" machine has a similar meaning.

With the detector finally constructed on stripboard, this allowed the unit to be put through its paces in the real world. In the UK it is well known by detectorists that field detecting requires a detector that is very good at rejecting iron due to the amount of ferrous trash in the ground. Some areas are worse than others, and my test garden is particularly bad. If you were to use a PI and dig all targets, there would be no lawn left as there is a signal every few inches. The test garden has 5 English pennies buried there (the smaller new penny which is a similar size to a U.S. cent, and not the much larger Victorian and Georgian ones) all of the same date, and without an iron core. The first is buried at 2", the next at 3", up to 6". Each one was epoxied to the end of a length of dowel and driven into the ground. This means that, even after many years, the coins remain at the same depth, and there is nothing directly above them except the wooden dowel. Virtually all the detectors that have been tested here can find the first coin, although some find it easier than others. Only a Viking 5 failed to find any of the coins. This is because it is a non-motion detector, and is highly affected by ground response. Although the Viking performs admirably on the bench, it requires constant adjustment of the threshold control when used in the real world. The majority of decent detectors can find the second coin, but none are able to detect the last three. However (on a good day) and knowing the exact position of the third coin, you can sometimes fool yourself that you wouldn't have missed it on a real dig. So this is a very challenging test that is typical of searching farmland in the UK.

The Hammerhead design uses a self-adjusting threshold (SAT) to maintain stability and to avoid constant manual tweaks to the audio threshold level, whereas Raptor is a motion detector that uses a double-differentiating architecture for its GB and DISC channels, also not requiring constant threshold adjustments. So which method, or combination of methods, could we use for this hybrid detector?

At this point I could document the numerous variations of architecture, the hours of head scratching, and even longer periods of near despair that followed, in an attempt to discover the perfect combination. There were times when I became somewhat disillusioned with the idea, and began to think the whole concept might be fundamentally flawed. Some of the problems I attributed to the fact that the PI and DISC preamps were receiving signals from two loops of differing sizes, and surmised that it was possibly the overlap between the two which allowed chatter to occur, either that or there was some imbalance between the two channels. There were also many headaches associated with generating the required timing pulses using the PIC's embedded software, as several of these pulses overlapped, and it was a nightmare to make these adjustable from the menu system. The detector also became unstable at random intervals for no obvious reason. Many different combinations of non-motion, single and double differentiating filters were employed, but the best results were always obtained with double-differentiating. However, double-differentiating the broadband PI signal was a major source of noise.

Eventually after much effort, the noise, instability, chatter, and timing issues were all resolved. Hence the remainder of this book documents the *final* working version, but with short references to the various problems that needed to be solved.

How does it work?

Although not intended for operation with a PI, a Troy coil was used for all the initial testing. This is a concentric coil with a TX diameter of 22cm and RX diameter of 8.5cm. The inductance values were 1.12mH and 6.95mH respectively.

With reference to Figure 2-2: The outer TX coil was used as a mono, being driven by the PI TX circuit, and the usual decay curve was fed to a preamp in the normal way. The output of the preamp was sampled and fed to a filter channel that double-differentiated the received signal in an identical manner to many VLF designs. There was also a second sample pulse that occurred much later in the TX cycle, which was subtracted from the received signal to provide Earth field elimination (EFE). At this point the design was a classic PI detector except that it was motion based. The usual damping resistor was connected across the TX coil to achieve critical damping.

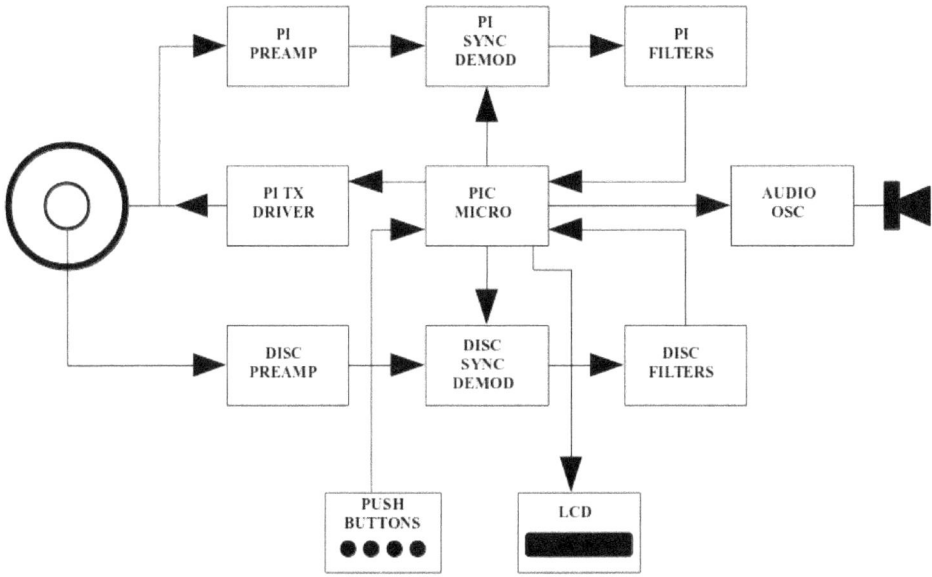

Fig. 2-2: Block Diagram of Detector

The RX coil was resonant tuned with a 33nF capacitor to a frequency of approximately 10.5kHz, resulting in a decaying sine-wave that was fed to a second preamp to provide ferrous/non-ferrous discrimination. A DISC sample pulse was positioned on the RX waveform such that non-ferrous targets gave an amplitude change in one direction, and ferrous targets gave a change in the other direction. Adjusting the DISC pot to a higher setting eliminated other non-ferrous items (such as foil and pulltabs) in exactly the same way as a VLF. This was interesting, because there was now a combination of two separate methods of discrimination. Firstly, the PI channel could eliminate targets based on conductivity, and the DISC channel could provide rejection based on phase.

 **Resonant PI Patent**

It was around the time of this testing that I became aware of a new patent for a resonant PI that describes a similar method of using a resonant coil, but in this case all the signal processing is done on the RX signal only, whereas the Voodoo Project has two separate preamps, one for the PI channel, and the other for the DISC channel. Also, the general concept of making the RX coil resonant at a specific frequency is already well known in the art, and has been included in many other detectors from companies such as Heathkit and Radio Shack. The TX coil is not measured in any way, and for some reason I do not comprehend, the sync demod reference is not taken from the TX oscillator. Instead it is derived by pulsing the RX coil on alternate cycles to create a reference signal. This appears to be an unnecessary complication, as the TX oscillator can be used as the reference without any problems. However, it is one of the novel claims in the patent. By only measuring the signal present on the RX coil and extracting the in-phase and quadrature components, in the same way as a standard VLF, means the detector will require ground balancing. This particular patent therefore appears to be describing a VLF detector driven by a PI TX circuit.

According to the patent: "*In a PI detector, high voltage back electromotive force, back EMF, commonly referred to as a flyback pulse, follows the termination of each transmit pulse. The energy in this pulse is not used by pulse induction metal detectors and is thus wasted.*"

If I am reading this correctly, this is a huge misconception, as it is actually the collapsing magnetic field (which occurs during the flyback pulse) that generates eddy currents in any nearly targets. Many people believe that the targets are being *charged* during the on-time, and that the off-time does nothing. This is patently incorrect (excuse the pun).

The detector being referred to in the patent is actually the Pulse Devil, designed by David Emery, and to all accounts it works very well and detects targets at an impressive depth. At least that is the conclusion from the small number of people who have witnessed it in operation. This may be so, because it is actually a VLF detector with a very powerful PI transmitter (IMHO).

In the following chapters, the various functions performed by the electronic circuitry will be covered in detail, and the intricacies of the PIC's embedded code will be revealed.

Chapter 3

Detector Electronics

"Think left and think right and think low and think high. Oh the thinks you can think up if only you try!"

--- Dr. Seuss

Power Supply

As mentioned briefly in Chapter 2, the original version suffered some instability problems. These were traced to the power supply circuit and were caused by synchronization of the voltage converter to the TX oscillator in order to reduce noise. Since the power supply arrangement was copied from the original Hammerhead circuit it used an ICL7660A to generate the positive power supply rail. However, it appears that the TX oscillator frequency of 1000pps was too low for the ICL7660A, and as a result the extra opamps and the PIC required in this design were causing some loading issues. Although the voltage converter is capable of free-running using its internal 10kHz oscillator, this can cause unsynchronized voltage spikes on the input signal at the PI preamp. To remove this unwanted noise it is necessary to operate the voltage converter from an external oscillator that is synchronized with the transmitter.

In addition, the ICL7660A has a maximum operating voltage of 12V, and the battery pack is made up of 10x NiMH rechargeable AA cells which have a total voltage of 12V nominal. These cells (after charging) can reach around 1.35V in which case the battery pack voltage can become greater than the maximum voltage of the converter. According to the datasheet the *absolute* maximum rating is 13V, with the result that you might get away with it in practice for some time, but the subsequent stressing of the voltage converter will make it fail eventually. From personal experience the ICL7660A has a tendency to release the magic smoke if you as much as look at it the wrong way.

The LT1054, on the other hand, has a maximum operating voltage of 15V and is pin-compatible with the ICL7660A. The LT1054 can also operate with a much lower synchronization frequency, and was used as a replacement for the ICL7660A.

 **Magic Smoke**

The idea of "magic smoke" is a joke often made by electronics engineers. Electronic components operate as long as the magic smoke is contained within the device. If the smoke escapes, the component ceases to function. The conclusion being that the magic smoke is an essential part of its operation.

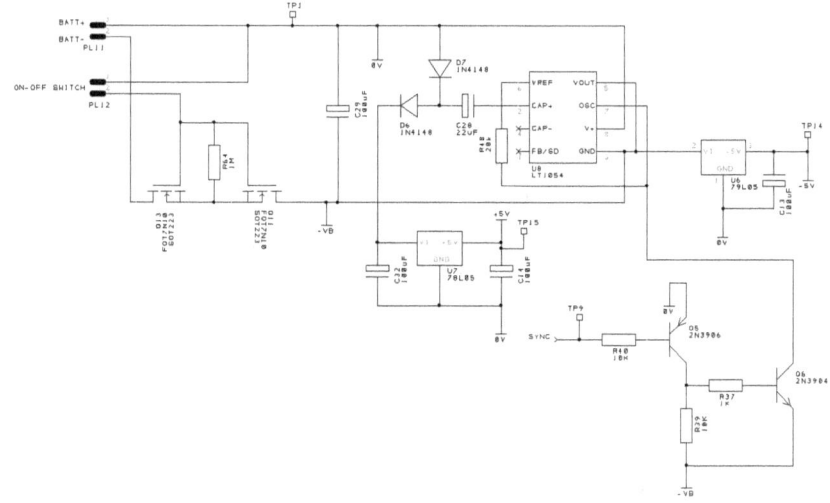

Fig. 3-1: Power Supply

One horrendous event that seems to happen to a lot of DIYers, is accidentally connecting the power supply or battery pack in reverse. Numerous times over the years people have posted in the Geotech forum that they've mis-connected the battery. Now their detector is dead, and what should they do? Basically, the only thing you *can* do is to cry a bit before grabbing a multimeter and a scope, and starting the business of fault-finding. Don't worry, as we've all done it at one time or another in the past. One simple solution to stop this happening is to use connectors that are keyed, or a 3-pin connector so that it can be fitted either way round. Although this still hasn't stopped some determined souls from managing to get the connections back-to-front. Some designs have a blocking diode in the supply line, but this causes a volt drop and wastes energy as heat in the diode. Sometimes a diode is placed across the battery input in reverse bias. The idea being that a reverse polarity supply will blow an inline fuse. Unfortunately the fuse is often missed out with the result that the diode lets out its magic smoke, along with some other components, and a section of PCB tracking gets burnt out just to add to the misery.

For Voodoo we're going to take a alternative fail-safe approach. Figure 3-1 includes a protection circuit that uses two FQT7N10 mosfets in series inserted in the negative voltage line. Such an arrangement only allows current to flow in one direction. If the battery pack is connected in reverse, current flow is blocked, and any sadness is averted.

It is important to note that Voodoo uses the battery positive as ground (0V), and this might cause some confusion until you can get your head around it. The battery pack consists of 10x AA rechargeable NiMH cells. (I recommend Duracell Recharge Ultra 2500mAh). The 79L05 is a linear voltage regulator which runs directly from the supply to provide the -5V rail. The +5V is generated by first using an LT1054 voltage converter configured as a voltage doubler to boost the supply to +24V with reference to the battery negative line. This in effect creates a +12V rail when referenced to battery positive (0V). A 78L05 linear voltage regulator is then used to convert this to +5V. The bipolar transistors and associated passive components are used to synchronize the LT1054 with the TX oscillator. This helps prevent any voltage converter switching transients from affecting the RX signals.

Apart from the mosfets used for reverse polarity protection, the PCB for this *final* version was created as a through-hole layout to make it easier to construct. SMD layouts are great for a design you plan to manufacture in large volumes, but not so good for a DIY job at home, especially when you're still debugging the thing.

PI Transmitter

The transmit oscillator pulses are generated by the PIC and used to drive the mosfet (see Figure 3-2).

The PIC generates a signal (TX_OSC) to pulse the mosfet at defined intervals. When TX_OSC is high, the mosfet is turned on *passively* by the pullup resistor R12, causing current to flow in the TX loop. When TX_OSC goes low, the mosfet is *actively* turned off by Q2. The series diode is there to isolate the mosfet capacitance (Coss) during flyback, and a detailed explanation of the function of this diode is provided in Appendix D.

This latest version uses active mosfet turn-off to provide a higher flyback voltage, but being careful not to drive the mosfet into avalanche mode. The faster the mosfet turns off the faster the decay curve, which allows sampling to occur earlier than if passive turn-off were used.

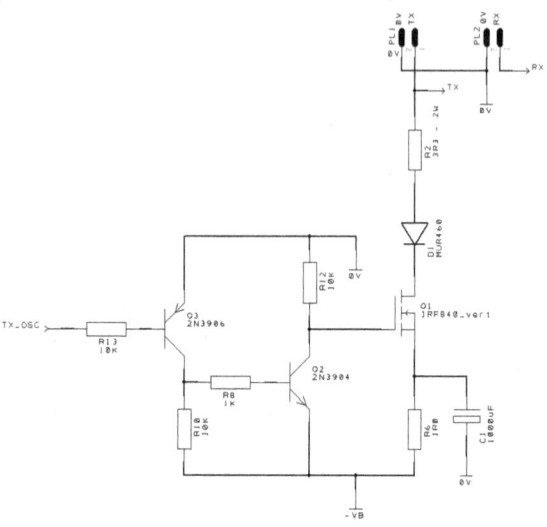

Fig. 3-2: Transmitter (TX) Circuit

There is a 3R3 resistor in series with the TX loop and the mosfet, which limits the maximum current. If the nominal voltage of the battery pack is assumed to be 12V, and the TX loop resistance is 4R4, then a quick hand calculation indicates the maximum possible current (if allowed to flat-top) will be:

$$I_{max} = \frac{V}{R_T} = \frac{12}{7.7} = 1.56A$$

where $R_T = 4.4 + 3.3 = 7.7$ ohms.

For more information on the subject of flat-topping, see Appendix C.

If the TX pulse width is 150µs, the predicted peak current is:

$$I_{peak} = I_{max}[1 - exp(\frac{-R_T * t}{L})] = 1.56 * [1 - exp(\frac{-7.7 * 150\mu}{1.12m})] = 1.00A$$

The wider the TX pulse width the less detecting time will be possible as the battery pack will be depleted more quickly. For example, a TX pulse width of 150μs gives an average current consumption of 235mA.

Therefore:

$$Detecting\ time = \frac{battery\ pack\ capacity}{average\ current\ consumption} = \frac{2500mAh}{235mA} = 10.64hrs$$

In practice the peak current was measured as 853mA, which is considerably lower than expected from hand calculation. Of course the above calculations do not take into account the resistance of the series diode or the mosfet's R_{DS}.

The energy stored in the TX loop just prior to switch-off is then:

$$E = \frac{1}{2}LI^2 = \frac{1}{2} * 1.12m * (853m)^2 = 407.5\mu J$$

The combination of C1 (1000μF) and R6 (1R0) effectively decouples the mosfet from the battery supply, preventing excessive ripple on the input supply rail. In practice this ripple has little effect on either the +5V or -5V supplies, as these are isolated from the supply rail by the linear voltage regulators and associated smoothing capacitors.

PI Preamp

Hammerhead used a single stage preamp with a gain of 1000x. This has been replaced in the Voodoo design by a dual-stage circuit having two opamps in series, each with a gain of 33x, giving a total gain of 1089x. (See Figure 3-3.) The idea here was not necessarily to provide early sampling per se, but to allow a shorter main sampling delay to be used than was previously possible with the 1.12mH TX loop.

To add some flexibility to the hardware when testing different coil configurations, the second stage of the PI preamp includes two sets of jumpers (PL7 and PL8) that allow it to be configured as either inverting or non-inverting. For the Troy coil, PL7 pins 1 and 2 are connected, and PL8 pins 2 and 3 are connected. In this case the second stage is configured in non-inverting mode. Since the PI preamp needs to be able to amplify some very small signals, the supply lines should be kept as clean as possible. The components R32/C9 and R17/C6 act as low-pass noise filters for the +5V and -5V lines respectively. Trimmer R30 is the null offset adjustment. This generic arrangement is similar to that used by Hammerhead, which applies an offset current to the inverting pin of the first stage via R18.

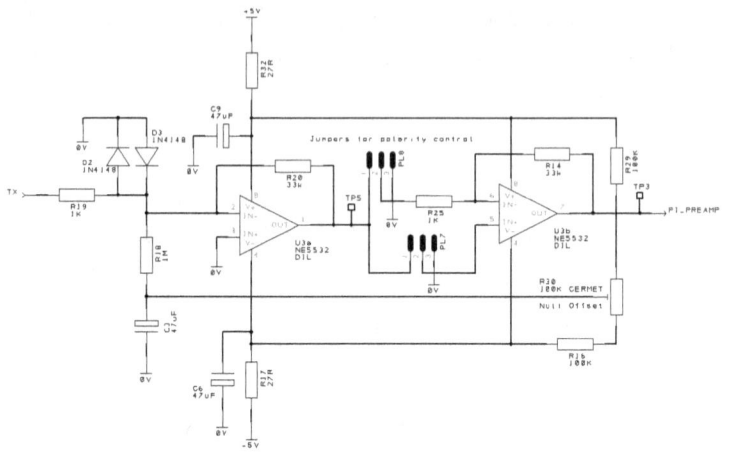

Fig. 3-3: PI Dual-stage Preamp

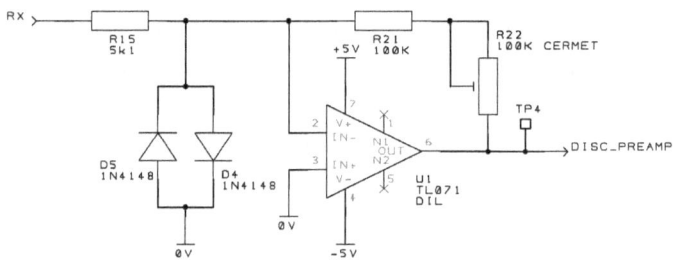

Fig. 3-4: DISC Single-stage Preamp

DISC Preamp

In an attempt to keep both the PI and DISC channels the same the DISC preamp also had a gain of 1000x in the prototype designs, but this was found to be far too high for stable operation. In fact it only needed to have the same gain as the Raptor VLF detector, which was 19.6x. Since the DISC preamp is connected in an inverting amplifier configuration, the jumpers in the PI preamp can be set to ensure that both outputs have the same polarity if this is found to differ between coils. The importance of this will become evident later when we examine the operation of the embedded software. For flexibility during experimentation, however, the trimmer (R22) has been included to allow the gain to be increased if required. (See Fig. 3-4)

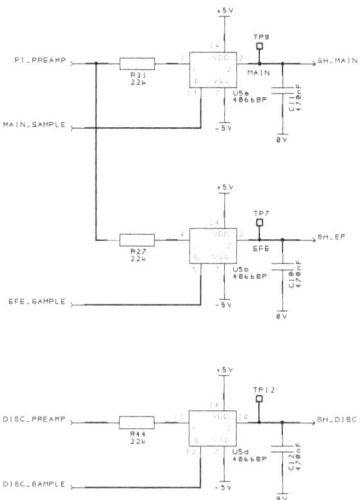

Fig. 3-5: PI and DISC Synchronous Demodulators

Synchronous Demodulation

Both the PI and the DISC channels need to be demodulated to extract the target signal. (See Fig. 3-5) The PI preamp output is fed into a CMOS bilateral switch (U5a) which samples the signal after the mosfet is turned off and the opamp comes out of saturation. This is called the main sample pulse, and the time between mosfet switch-off and the main sample is called the main sample delay. The bilateral switch acts like a sample-and-hold, with R31 and C11 providing some lowpass filtering. There is also a secondary sample taken (U5b) much later in the TX period. Due to the huge gain of the PI preamp the detector will produce a response as the coil is moved through the Earth's magnetic field by so-called "flux cutting". Since the signal due to the Earth field (EF) will be present (and substantially unchanged) in both the main and secondary samples, a simple subtraction of the secondary from the main will result in only the target signal remaining. This is commonly referred to as Earth field elimination (EFE). R27 and C10 provide sample-and-hold plus lowpass filtering for the EF signal.

The DISC sample pulse is positioned on the second half-cycle of the DISC preamp's output signal for the Troy coil that was used during testing . The correct positioning of the sample pulse is important, as the DC level (as measured at the output of the bilateral switch) must decrease in amplitude for non-ferrous targets, and increase for ferrous. (See Figs. 3-6 to 3-8)

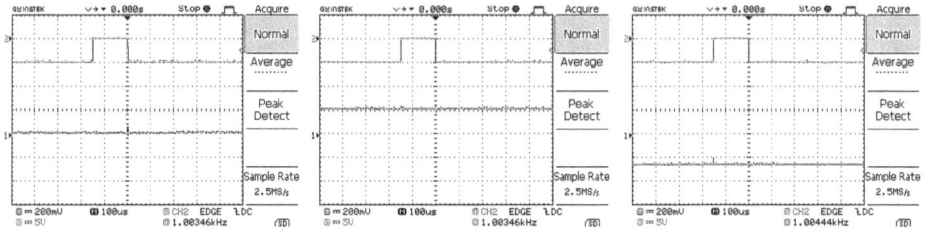

Fig. 3-6: No target　　　**Fig. 3-7: Non-ferrous**　　　**Fig. 3-8: Ferrous**

PI and DISC Channel Filters

One of the primary problems experienced with the original prototypes was some unwanted chatter in the presence of iron. Because the PI preamp is essentially broadband by nature, I was concerned that the use of a double-differentiating architecture for the filters may be the cause. Other prototypes were built that allowed the filter channels to be configured as either single or double-differentiating and could be used to test out various theories. After much experimentation, it was discovered that double-differentiating still gave the best results when combined with the sample-and-hold gates, and some *tweaking* with software. Interestingly, the combination of double-diff for the PI channel, and single-diff for the DISC channel gave excellent results on the bench. Unfortunately in the test garden, it was completely hopeless and unable to detect the first coin, even at 2 inches depth.

Since Voodoo is designed to be a motion detector, it will respond to the rate-of-change of the received signal rather than the signal itself. This allows it to ignore any slow moving variations such as those from ground signal changes, or component and temperature drift. The filter circuits are therefore based around differentiators which provide an output that is proportional to the derivative of the input. The filters used in Voodoo are known as "practical differentiators" because they introduce a cutoff frequency that is below the upper cutoff frequency of the opamp. (See Figs. 3-9 and 3-10) Without this, there would be a lot of high frequency noise at the output.

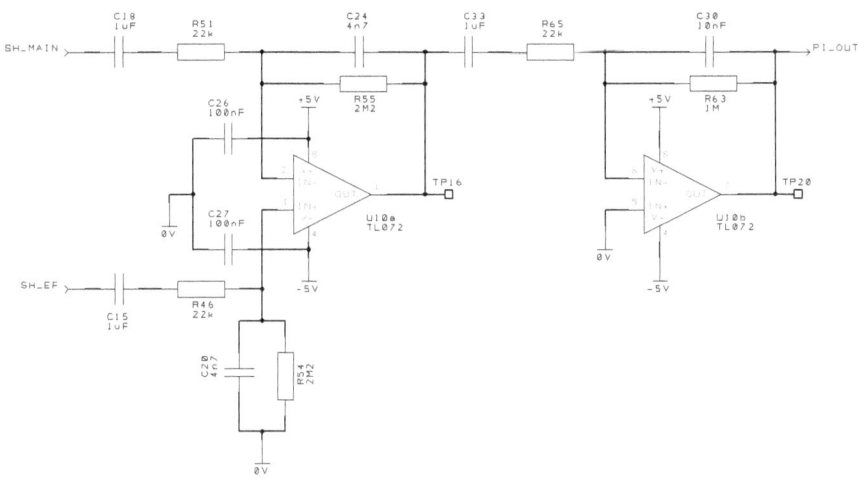

Fig. 3-9: PI Channel Filters

In the low frequency region where differentiation occurs, the gain ramps up from DC (0Hz) at a rate of 20dB per decade with a break frequency close to 7Hz.

$$f_b = \frac{1}{2\pi RC} = \frac{1}{2 * \pi * 22k * 1u} = 7.2Hz$$

The upper cutoff frequency, close to 15Hz, was chosen to reject both the mains frequency (50Hz or 60Hz, depending on where you live) and the transmit pulse repetition rate of 1kHz.

$$f_c = \frac{1}{2\pi RC} = \frac{1}{2 * \pi * 2.2M * 4.7n} = 15.4Hz$$

The filter section is a double-differentiator, and therefore a second differentiation stage is required. The break frequency is the same as the first stage, but the cutoff frequency is closer to 16Hz.

$$f_c = \frac{1}{2\pi RC} = \frac{1}{2 * \pi * 1M * 10n} = 15.9Hz$$

Detector Electronics

There is a slight difference between the two cutoff frequencies because the two stages have different gains, and we are restricted to using standard component values.

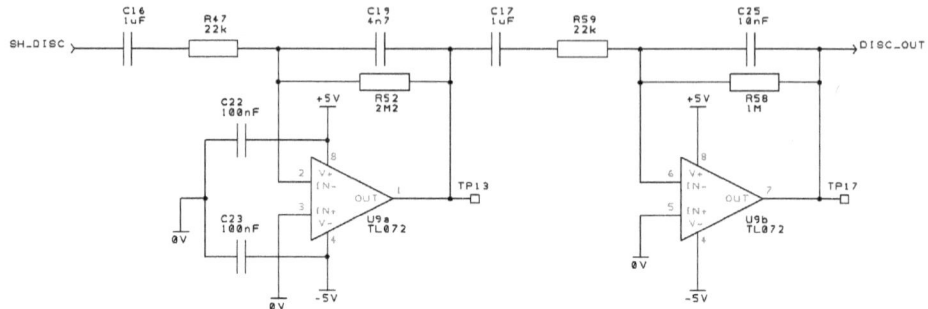

Fig. 3-10: DISC Channel Filters

The DISC channel filters are exactly the same as the PI channel, with the exception that the first stage is a straightforward inverting configuration, as there is no requirement to subtract an EF signal.

Audio Oscillator

The audio frequency is generated by a common 555 timer. Pins 2 (trigger) and 6 (threshold) are connected together so that it will trigger itself and run as an astable multivibrator. The capacitor C35 charges and discharges between one-third and two-thirds of VCC. The charge and discharge times are independent of the supply voltage, as is the oscillation frequency.

The charge time (output high) is given by:

$$t_1 = 0.68(R_1 + R_2)C_1$$

and the discharge time (output low) by:

$$t_2 = 0.68 \ R_2 \ C_1$$

Thus the oscillation frequency is given by:

$$f = \frac{1.46}{(R_1 + 2 \ R_2)C_1}$$

where R_1, R_2 and C_1 are equal to the values of R67, R68 and C35 in Figure 3-11.

Therefore we have:

$$t_1 = 1.50ms, t_2 = 0.82ms, and f \approx 430Hz$$

The 555 timer output is fed to a simple transistor audio stage to drive either a loudspeaker or headphones, and the PIC uses an open-drain output (RA4) to create a logical AND function that disables the audio when connected across Q10.

PIC Microcontroller

Last, but certainly not least, is the PIC microcontroller. In the original prototype a PIC16F877A was used, as this was supplied with the EasyPIC development system from Mikroelektronika, and was programmed using mikroBasic. This *final* version uses a PIC18F4520, and was programmed using mikroC Pro for PIC.

The PIC is configured for 20MHz operation using an external crystal. It is important that this is a parallel cut crystal, as a series cut crystal will result in an operating frequency outside of the manufacturer's specification The oscillator type must also be configured for high-speed crystal/resonator (HS) operation.

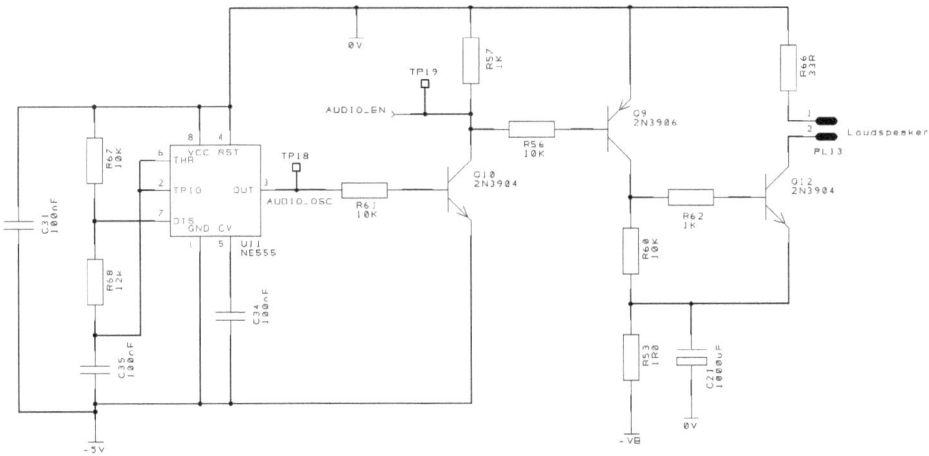

Fig. 3-11: Audio Oscillator

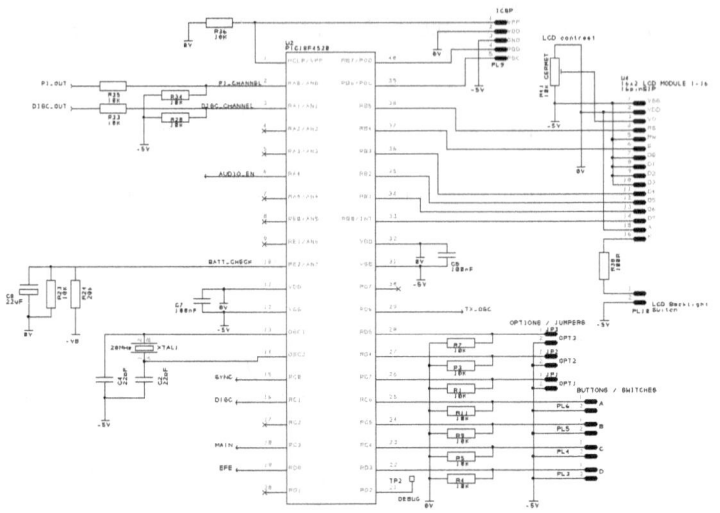

Fig. 3-12: PIC Microcontroller

With reference to Figure 3-12:

The PCB incorporates an in-circuit system programming (ICSP) interface to allow hex code to be loaded into the internal flash memory without having to remove the PIC from the board. A Microchip MPLAB PICkit4 was used to perform this operation.

There are 13 10-bit analog-to-digital converter (ADC) channels available, but this project only requires the use of 3 ADC channels:

AN0: Reads the analog output signal from the PI channel.

AN1: Reads the analog output signal from the DISC channel.

AN7: Battery pack voltage measurement.

Since the PI and DISC channel outputs swing between +5V and -5V and the PIC has VDD connected to 0V and VSS connected to -5V, resistor dividers are used at the inputs of AN0 and AN1 to persuade the voltage to swing between 0V and -5V.

A voltage divider is also present at the input of AN7, such that the input voltage will be one third of the battery pack voltage. The ADC can measure up to 5V, so the maximum allowed battery pack voltage is limited to 15V. This provides a safety margin, as with 10x AA rechargeable NiMH batteries the maximum is not expected to exceed 13.5V.

Port RA4 has an open-drain output, and is used to disable the audio oscillator by pulling down the collector of Q10.

Port RC0 outputs a synchronization pulse to act as an external oscillator for the LT1054 voltage converter. This pulse is generated under software control, and must be present for the +5V power rail to exist.

Ports RC3, RD0, and RC1 are the sample pulses for the main, EFE and DISC bilateral switches respectively. Since the PIC is connected between 0V and -5V, it is necessary to perform a voltage level conversion on each of these signals such that they swing between +5V and -5V. This is accomplished by the circuitry shown in Figure 3-13.

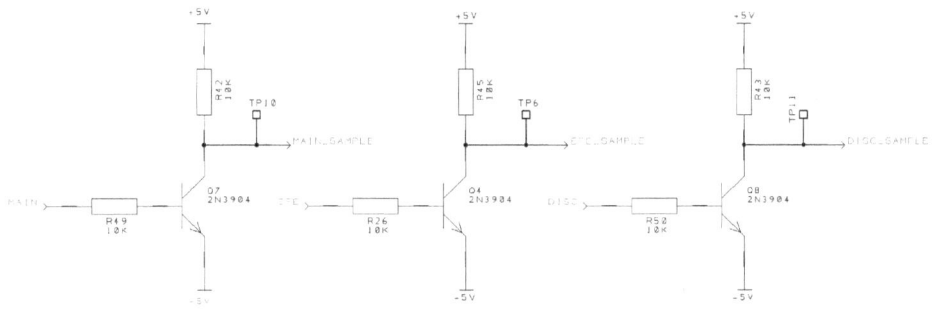

Fig. 3-13: Voltage Level Converters

Port RD2 generates a *debug* pulse that is useful for synchronizing an oscilloscope when taking various measurements. For example, the *debug* pulse is the upper trace (Channel 2) in Figs. 3-6 to 3-8.

There are 4 pushbuttons on the front panel of the detector that are connected as follows:

Menu (RC6), **Up** (RC5), **Down** (RC4), and **Enter** (RD3)

The function of each of these buttons is explained in detail in the Embedded Software chapter. When a button is pressed, the respective digital input on the PIC is pulled low. Otherwise all the button inputs are pulled high by a 10k resistor. There are also 3 hardware jumpers that could potentially be used to enable various features, but these are not currently used by the software.

Port RD6 outputs the transmit oscillator signal that switches the TX mosfet on and off.

Ports RB0 to RB5 are used to drive a 2-line 16-character LCD module in 4-wire display mode. There is an option for an external switch to turn on/off the LCD backlight. However, the White's MXT display (used by the version discussed in this book) did not possess a backlight. A more standard 16 character display could easily be substituted without any changes, but the MXT display was already available, and the characters on the screen are larger than the standard displays. All of these types of LCD require the contrast to be correctly set, and an onboard trimmer (R41) can be adjusted to provide a clear display.

Miscellaneous

There is (of course) a resistor (R_d) required across the coil's outer (PI) loop to achieve critical damping. This damping resistor was connected to the rear of the coil connector inside the electronics enclosure, and a wire link was added to the coil plug wiring to connect this resistor into circuit. This allows a custom-built coil, requiring a different value of damping resistor, to be connected by placing the resistor inside the coil housing and leaving out the wire link. Alternatively if putting a resistor inside the coil housing is not desirable, a larger value damping resistor could be fitted to the rear of the coil connector, and a low wattage parallel resistor put inside the coil plug to obtain the correct value for critical damping.

A capacitor is also required across the coil's inner (DISC) loop to create a decaying sine wave of the desired frequency. In this case, the combination of the Troy RX loop inductance of 6.96mH and a parallel capacitor of 33nF resulted in a frequency of 10.5kHz, as seen in Fig. 2-1. The 33nF capacitor was placed inside the coil plug.

Fig. 3-14 shows the coil socket wiring, and Fig. 3-15 shows the coil plug wiring that was used in this project.

Testpoints

There are 20 testpoints provided on the PCB to allow measurements to be made easily.

TP1 is the positive terminal of the battery pack, and is used as the reference point (0V) for all other measurements.

TP2 is the *debug* signal, which tracks the period of the TX signal. This signal is used to trigger the oscilloscope (channel 2) so that delay and pulse width measurements may be acquired. In the scope images below, the *debug* signal is the top trace (channel 2), except for the audio oscillator, main, EF, and DISC sample images.

TP3 is the output of the second stage of the PI preamp. It is this signal that is sampled by the main and EF sync demods (Fig. 3-17).

TP4 is the output of the DISC preamp. Unlike the PI preamp, this is a single-stage amplifier, and is the signal that is sampled by the DISC sync demod (Fig. 3-18).

TP5 is the output of the first stage of the PI preamp. Hammerhead used a single-stage preamp, but the use of a dual-stage amplifier allows the signal to settle more quickly, which certainly helps with the Troy coil (Fig. 3-18).

TP6 is the Earth field sample that is taken much later in the TX period. The main and EF sample pulse widths need to be identical in order for the magnetic field to be cancelled (Fig. 3-20).

TP7 and **TP8** respectively are the EF and main outputs from the sync demods. The demodulated main signal amplitude increases in the presence of a target, regardless of whether it is non-ferrous or ferrous. If the target is highly conductive and physically large, there may be some residual target signal present in the demodulated EF sample. If that is the case, this will also increase in amplitude for all targets.

TP9 is the synchronization pulse for the LT1054 voltage converter (Fig. 3-22).

TP10 is the main sample pulse for the sync demod. The pulse starts after the preamp comes out of saturation (Fig. 3-19).

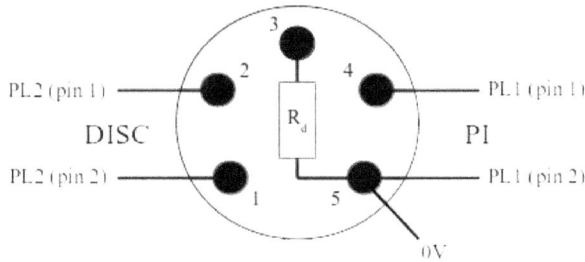

Fig. 3-14: Coil Socket Wiring (rear view)

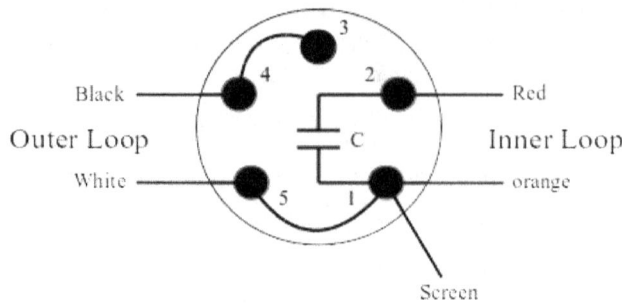

Fig. 3-15: Coil Plug Wiring (rear view)

TP11 is the DISC sample pulse for the sync demod. The pulse is positioned on the second half-cycle of the decaying sine wave (Fig. 3-21).

TP12 is the output from the DISC sync demod, whose amplitude changes in different directions depending on whether the target is non-ferrous or ferrous (Figs. 3-6 to 3-8).

TP13 is the output of the first stage of the DISC filter channel. This filter differentiates the output signal from the DISC sync demod, resulting in a signal that changes depending on the rate-of-change of the input.

TP14 is the negative 5V supply line (-5V).

TP15 is the positive 5V supply line (+5V).

TP16 is the output of the first stage of the PI filter channel. This filter differentiates the output signal from the PI sync demod, resulting in a signal that changes depending on the rate-of-change of the input. It contains only the signal from the target, after the EF signal has been subtracted.

TP17 is the second stage of the DISC filter channel. This filter differentiates the output signal from the first stage, so that the output is the rate-of-change of the-rate-of-change of the signal from the DISC sync demod. In other words, the original input signal is double-differentiated.

TP18 is the output from the audio oscillator (Fig. 3-23).

TP19 is the audio enable signal from the PIC.

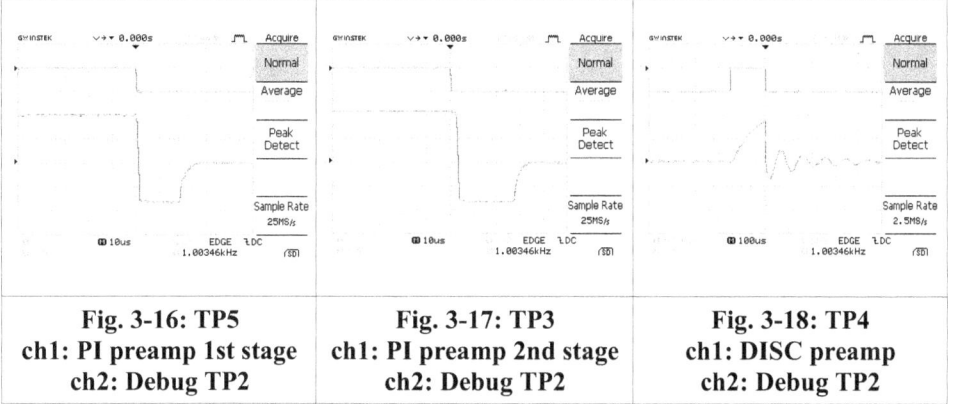

Fig. 3-16: TP5	**Fig. 3-17: TP3**	**Fig. 3-18: TP4**
ch1: PI preamp 1st stage	**ch1: PI preamp 2nd stage**	**ch1: DISC preamp**
ch2: Debug TP2	**ch2: Debug TP2**	**ch2: Debug TP2**

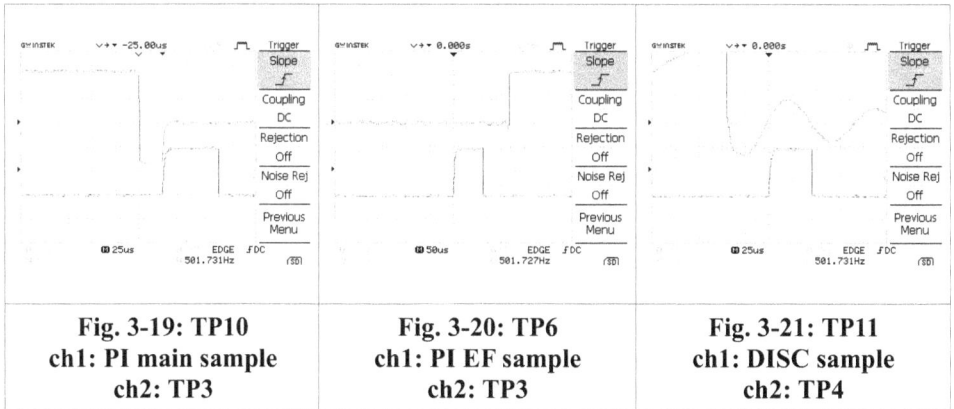

Fig. 3-19: TP10	**Fig. 3-20: TP6**	**Fig. 3-21: TP11**
ch1: PI main sample	**ch1: PI EF sample**	**ch1: DISC sample**
ch2: TP3	**ch2: TP3**	**ch2: TP4**

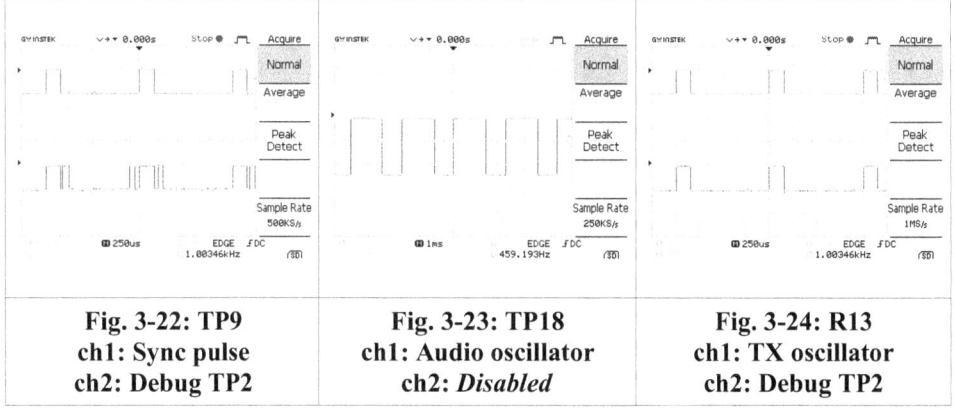

Fig. 3-22: TP9	**Fig. 3-23: TP18**	**Fig. 3-24: R13**
ch1: Sync pulse	**ch1: Audio oscillator**	**ch1: TX oscillator**
ch2: Debug TP2	**ch2: *Disabled***	**ch2: Debug TP2**

TP20 is the second stage of the PI filter channel. This filter differentiates the output signal from the first stage, so that the output is the rate-of-change of the-rate-of-change of the signal from the PI sync demod. In other words, the original input signal is double-differentiated.

Note: There is no testpoint for the TX oscillator as this is identical to the *debug* signal.

The locations of the 20 testpoints on the PCB are shown Fig. 3-25, where they have been highlighted as black squares.

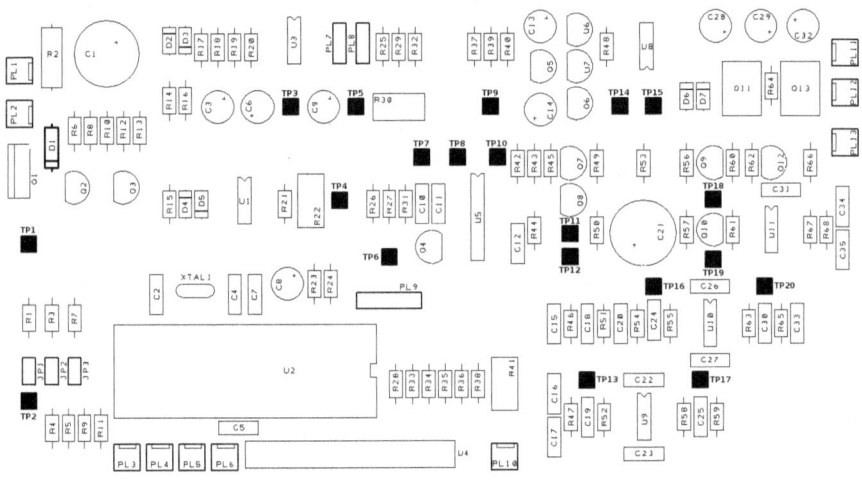

Fig. 3-25: Location of Testpoints on PCB

Chapter 4   *Detector Hardware*

"Science is about knowing; engineering is about doing."

--- Henry Petroski

Sadly, many homemade detector projects never make it off the bench and into the real world for testing. If they do, then they're often constructed using plastic piping for the stem, sections of gutter for the arm cup, and frisbees for coil shells.

Nowadays there is no excuse for bodging the engineering side of a project, as coil shells are readily available on the Internet, as are professionally made stems, arm cups, and all the fittings. All of these can be acquired at quite reasonable prices.

For the Voodoo project, a White's Electronics MXT 3-section straight stem, with arm cup and control panel / display, were used for most of the construction, and the PCB and battery pack were contained in a standard plastic enclosure. Previous prototype versions of Voodoo used aluminium enclosures, but these were found to make the detector very heavy to use over long periods. Although it was feared that external interference may become a problem if a plastic enclosure was used, this does not appear to have been the case in practice.

Battery Pack

As mentioned in Chapter 3 (Detector Electronics), the battery pack consists of 10x AA NiMH batteries. The particular ones used in this project are Duracell Recharge Ultra with a capacity of 2500mAh. In order to contain the batteries securely within the enclosure, a 3D printed fixture was designed using FreeCAD, and an STL file generated. Fig. 4-1 shows the battery box as viewed in FreeCAD. When inserted into the electronics enclosure, the battery box was turned upside down so that the batteries are firmly secured against the base (Fig. 4-2).

The STL file was read into Repetier-Host for examination, and CuraEngine was used to slice the model and generate a GCode file.

The GCode file was then transferred to a Geeetech i3 Pro B 3D printer, and the battery box was printed using PLA filament.

The finished battery box (with batteries fitted) is shown in Fig. 4-3.

STL File

STL is either short for Stereolithography, **S**tandard **T**riangle **L**anguage, or Standard **T**essellation **L**anguage. Nobody really knows, as its original meaning was forgotten long ago. Its purpose is to encode the surface geometry of a 3D object.

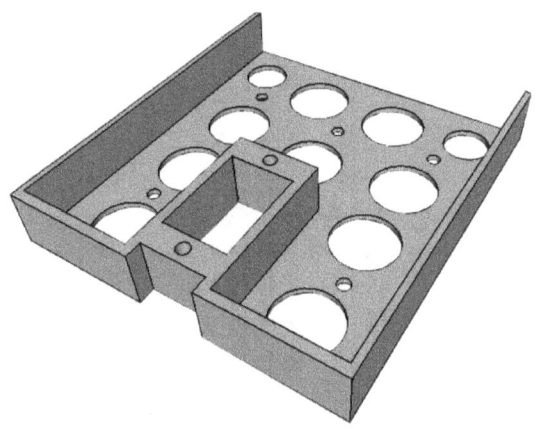

Fig. 4-1: Battery Box as Viewed in FreeCAD

Printed Circuit Board

The PCB is fitted inside the plastic enclosure above the battery pack mounted on spacers. This allows easy access to the onboard trimmers and the numerous testpoints (Fig. 3-25). Since all the external connections are via plug-in headers, removal of the PCB is a relatively easy process if access to the connectors or the battery pack is required. The PCB can be seen within the enclosure in Fig. 4-4.

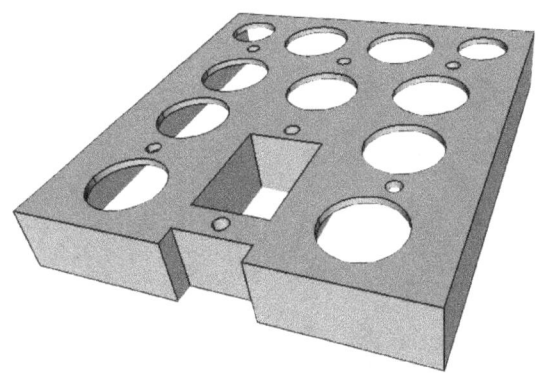

Fig. 4-2: Battery Box – upside down

Fig. 4-3: Completed Battery Box with Batteries Fitted

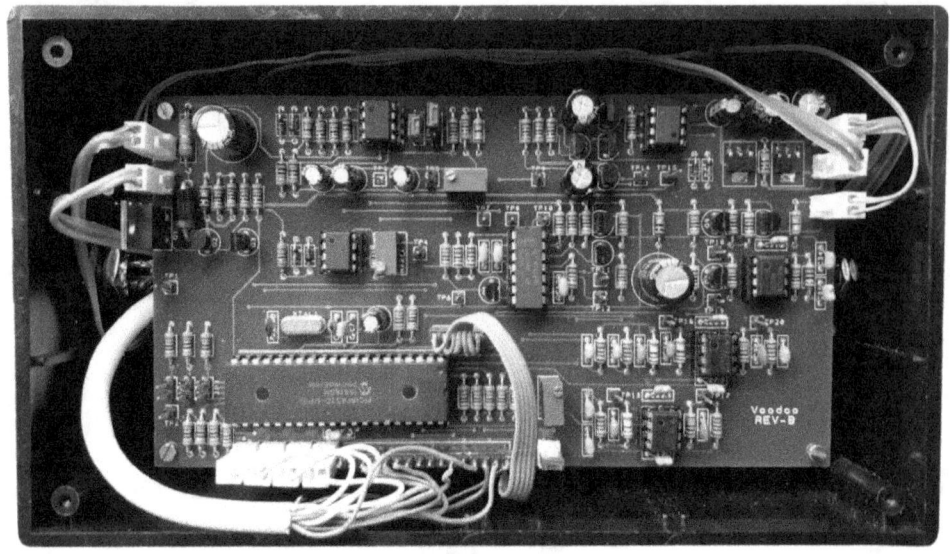

Fig. 4-4: PCB Installed in Enclosure

Battery Charger Socket

Since the power source of Voodoo consists of rechargeable NiMH batteries it stands to reason that a suitable charger is required, especially since the battery pack is not designed to be removed for the purposes of charging. A charger socket has therefore been fitted to the rear of the enclosure behind the arm cup (Fig. 4-5).

The design of a suitable charger is provided in Appendix B, together with a complete description of the circuit and a suitable PCB layout.

On-Off Switch

There is only one control located on the electronics enclosure, and that is the on-off switch (Fig. 4-6).

Fig. 4-5: Battery Pack Charger Socket

Fig. 4-6: Location of On-Off Switch

Coil Connector Socket

The wiring for the coil connector socket is shown in Chapter 3 (Fig. 3-14). The socket is a 5-pin connector, which is a fairly standard type used by several metal detector manufacturers. It is located at the front of the electronics enclosure (Fig. 4-7).

Fig. 4-7: Coil and Headphone sockets

Headphone Socket

The headphone socket (located on the same enclosure panel as the coil socket) is a standard 6.35mm jack (Fig. 4-7). Although an internal speaker was not fitted, this could easily be added if required, as the headphone socket is capable of automatically switching to headphones when they are plugged in.

ICSP Connector

The in-circuit system programming (ICSP) connector was made accessible without having to remove the cover of the enclosure by including a 5-pin socket on the side panel. The connector used is a physically smaller version of the coil connector. As can be seen in Fig. 4-8 and Fig. 4-9, this connector also includes a removable plastic cover. An MPLAB PICkit 4 was used to program the PIC via MPLAB's integrated programming environment (IPE).

Fig. 4-8: ICSP Socket – Covered

Fig. 4-9: ICSP Socket - Uncovered

Display and Control Buttons

The Voodoo embedded software requires an HD44780 compliant LCD controller, and it just so happens that the MXT control box (Fig. 4-10) came complete with a compatible LCD. The advantage being that the characters on the display are much larger than the standard size displays, and more easily seen by the operator. Of course, a standard compliant LCD could be used instead.

The control buttons are located below the display. These are fairly ordinary pushbuttons, that are mounted on a simple 3D printed panel. A 13-core cable provides the wiring connections for both the LCD and the pushbuttons. This cable was threaded down the handle and inside the stem until it reached the electronics enclosure.

The pushbuttons (from left to right) are Menu, Up, Down, and Enter. Their operation is described in Chapter 5 (Embedded Software).

Fig. 4-10: Display and Pushbuttons

Search Head

As described in previous chapters, the search head initially used to develop this project was a Troy 9" diameter concentric coil (Fig. 4-11).

Complete Detector

With the electronics enclosure, stem, arm cup, control box, search head, and other bits and pieces assembled, the complete detector can be see in all its glory in Fig. 4-12.

Fig. 4-11: Troy 9" Diameter Concentric Coil

Fig. 4-12: Complete Detector

Chapter 5   *Embedded Software*

"Always code as if the guy who ends up maintaining your code will be a violent psychopath who knows where you live."

--- John F. Woods (1991) – games programmer

The full [uninterrupted] source code is listed in Appendix G, and a detailed description of each section of the code is provided below in this chapter.

At the start of the source code there is some information about the project. The microcontroller is an PIC18F4520 with a 20MHz crystal, and the software was written using mikroC PRO for PIC from Mikroelektronika (www.mikroe.com). There is also a note stating that the sync pulse (which synchronizes the LT1054 voltage converter) must be present, otherwise there will be no +5V supply. Hence you can now see the reason why the PIC is powered with VDD connected to 0V, and VSS connected to -5V.

```
/*
 * Project name:
     Voodoo - PI/VLF Hybrid Metal Detector
 * Copyright:
     (c) George Overton, 2020.
 * Revision History:
     - V1.0 Initial release
 * Description:
     Pulse Induction / Continuous Wave Hybrid Metal Detector
 * Test configuration:
     MCU:              PIC18F4520
     Oscillator:       20 MHz Crystal
     Ext. Modules:     None.
     SW:               mikroC PRO for PIC
                       http://www.mikroe.com/mikroc/pic/
 * NOTES:
     Sync signal must be present for +5V supply to work.
*/
```

Next we need to define which ports are to be inputs, and which are to be outputs. The ports on the PIC are tri-state, and the input definitions use high (logic 1) for inputs, and low (logic 0) for outputs. Please refer to Fig. 3-12 to compare the definitions with the schematic.

```
// Assign port I/O macro definitions
#define TRISA_INIT 0x0F;
#define TRISB_INIT 0xC0;
#define TRISC_INIT 0xF0;
#define TRISD_INIT 0x38;
#define TRISE_INIT 0x06;
```

MikroC PRO contains a library that supports HD44780 compliant LCD controllers through the 4-bit interface. Before using this library it is necessary to define some global variables so that the library routines can understand how the LCD is physically connected. As we are using the 4-bit interface, only data bits 4 to 7 are used. Bits 0 to 3 are tied to the PIC's VSS line, which is -5V in this project.

```
// LCD assignments
sbit LCD_D4 at RB3_bit;              // Data line 4
sbit LCD_D5 at RB2_bit;              // Data line 5
sbit LCD_D6 at RB1_bit;              // Data line 6
sbit LCD_D7 at RB0_bit;              // Data line 7
sbit LCD_RS at RB5_bit;              // Register select
sbit LCD_EN at RB4_bit;              // Enable
```

The library also needs to know the direction of the tri-state lines that are connected to the LCD. These settings override the previous macro I/O definitions, although in this case there is no conflict between the two.

```
// LCD Port direction
sbit LCD_RS_Direction at TRISB5_bit;   // Register select
sbit LCD_EN_Direction at TRISB4_bit;   // Enable
sbit LCD_D7_Direction at TRISB0_bit;   // Data bit 7
sbit LCD_D6_Direction at TRISB1_bit;   // Data bit 6
sbit LCD_D5_Direction at TRISB2_bit;   // Data bit 5
sbit LCD_D4_Direction at TRISB3_bit;   // Data bit 4
```

A number of other port assignments are then required. Note that all writes should be made to the output latch (LAT) and all reads should be made to the port (i.e. the actual pin). This is important because there are situations where they could be different. The most obvious one is where the pin is accidentally shorted to ground. If you set the output latch high, the port will read as low. Usually this problem arises when the port has been incorrectly configured, and only reading and writing to the port (without using LAT) can result in hours of *fun* trying to figure out what's wrong.

The Voodoo Project

```
// Port assignments
sbit debug at LATD2_bit;                    // Debug output
sbit EN1 at LATD6_bit;                      // Enable MOSFET
sbit ps_sync at LATC0_bit;                  // Power supply sync pulse
sbit main_pulse at LATC3_bit;               // Main sample pulse
sbit efe_pulse at LATD0_bit;                // EFE sample pulse
sbit disc_pulse at LATC1_bit;               // DISC sample pulse
sbit audio_en at LATA4_bit;                 // Audio enable
sbit menu_btn_port at PORTC.B6;             // Keypad menu button
sbit up_btn_port at PORTC.B5;               // Keypad up button
sbit down_btn_port at PORTC.B4;             // Keypad down button
sbit enter_btn_port at PORTD.B3;            // Keypad enter button
```

So that we don't have to remember (or keep checking the schematic to find out) which
logic level turns something on or off, we will now set up some macro definitions to make
things easier. For example, instead of setting the output latch for RD6 high (logic 1) to
turn on the mosfet, we can use the macro definition "EN1_on" instead. Likewise we can
use "EN1_off" to turn off the mosfet. The use of EN1 for the mosfet enable is historical.
At one time there were two mosfets, but I won't get into that here.

```
// General macro definitions
#define EN1_on 1;                           // Turn on MOSFET
#define EN1_off 0;                          // Turn off MOSFET
#define main_on 0;                          // Turn on main sample
#define main_off 1;                         // Turn off main sample
#define efe_on 0;                           // Turn on EFE sample
#define efe_off 1;                          // Turn off EFE sample
#define disc_on 0;                          // Turn on DISC sample
#define disc_off 1;                         // Turn off DISC sample
#define audio_on 1;                         // Audio enabled
#define audio_off 0;                        // Audio disabled
```

A number of general constants are required by the program. These constants cannot be
changed during run-time.

```
// General Constants
const debounce = 100;                       // 100ms debounce time
const btn_off = 1;                          // Keypad button off
const btn_on = 0;                           // Keypad button on
const menu_btn = 1;                         // Keypad menu button
const up_btn = 2;                           // Keypad up button
const down_btn = 3;                         // Keypad down button
const enter_btn = 4;                        // Keypad enter button
const menu_active = 1;                      // Indicates menu is in use
const menu_inactive = 0;                    // Indicates menu is not being used
const pulse = 0;                            // Indicates pulse detection mode
const hybrid = 1;                           // Indicates hybrid detection mode
const blk_limit = 7;                        // Block display limit
```

The various timers within the PIC can be used to create accurate delays and pulse widths. However, due to software overhead, such as the time required to push data onto the stack and retrieve it when calling and exiting a function, some offset adjustments are required. Initially the offset values were set to zero, and the actual offset values were determined by measuring the extra delay on the oscilloscope.

```c
// Timer offset constants (us)
const txon_offset = 4;
const txpd_offset = 8;
const main_dly_offset = 5;
const main_smpl_offset = 4;
const efe_dly_offset = 10;
const disc_dly_offset = 9;
const disc_smpl_offset = 6;
```

Several variables are required by the program, and most of these are fairly obvious once you read the comments. Note that both the interrupt routine and the main section of code are controlled by a state machine.

```c
// Variables
unsigned i;                        // Generic variable
unsigned tmp;                      // Temporary variable
char int_state;                    // State machine for interrupt routine
char main_state;                   // State machine for main routine
char hybrid_cycle;                 // Hybrid operating cycle (0 = pulse, 1 = disc)
unsigned t;                        // Temporary storage for ADC readings
unsigned vbatt;                    // Battery voltage
unsigned pi_target;                // PI target voltage
unsigned disc_target;              // DISC target voltage
char loop_count;                   // Main loop count
unsigned pi_array[64];             // Running average array for PI target voltages
char pi_pointer;                   // Pointer into PI array
unsigned pi_tmp;                   // Temporary variable for PI array calculations
unsigned pi_total;                 // Total of pi_array contents
unsigned disc_array[64];           // Running average array for disc target voltages
char disc_pointer;                 // Pointer into disc array
unsigned disc_tmp;                 // Temporary variable for disc
unsigned disc_total;               // Total of disc_array contents
unsigned short pi_adc = 0;         // PI channel ADC
unsigned short disc_adc = 1;       // Disc channel ADC
unsigned short batt_adc = 7;       // Battery monitor ADC
char keypad_btn;                   // Keypad button that has been pressed
char result;                       // Records result of a button push
char menu_flag;                    // Flags state of menu system (active or inactive)
char menu_disp;                    // Indicates current menu display screen
char detect_mode;                  // Operating mode of (0 = pulse, else = hybrid)
unsigned pi_thr;                   // Audio threshold for PI channel
unsigned disc_thr;                 // Audio threshold for disc channel
char thr_disp;                     // Audio threshold for display purposes
char txon;                         // TX pulse width (us)
char txonh;                        // TX pulse width high byte
char txonl;                        // TX pulse width low byte
unsigned txpd;                     // TX pulse period (us)
char txpdh;                        // TX pulse period high byte
char txpdl;                        // TX pulse period low byte
char main_dly;                     // Main sample delay (us)
```

```c
char main_dlyh;          // Main sample delay high byte
char main_dlyl;          // Main sample delay low byte
char main_smpl;          // Main sample delay pulse width (us)
char main_smplh;         // Main sample pulse width high byte
char main_smpll;         // Main sample pulse width low byte
unsigned efe_dly;        // EFE sample delay (us)
char efe_dlyh;           // EFE sample delay high byte
char efe_dlyl;           // EFE sample delay low byte
char efe_smplh;          // EFE sample pulse width high byte
char efe_smpll;          // EFE sample pulse width low byte
char pi_num;             // Number of PI readings to average
char disc_dly;           // DISC sample delay (us)
char disc_dlyh;          // DISC sample delay high byte
char disc_dlyl;          // DISC sample delay low byte
char disc_smpl;          // DISC sample pulse width (us)
char disc_smplh;         // DISC sample pulse width high byte
char disc_smpll;         // DISC sample pulse width low byte
char disc_num;           // Number of disc readings to average
char accept;             // Accept (non-ferrous) counter
char accept_blk;         // Accept block display counter
char reject;             // Reject (ferrous)counter
char reject_blk;         // Reject block display counter
char update_disp;        // Update display flag
char meter_zero;         // Meter zero counter
char meter_zero_limit;   // Meter zero counter limit
char counter_limit;      // Accept and reject counter limit
```

The interrupt routine controls all the necessary timing by making extensive use of the PIC's internal timers (timer0 and timer1). The interrupt state machine (int_state) starts off at 0. In this state the mosfet is turned on, timer0 is loaded with the TX pulse width, and timer1 is loaded with the TX period. There are separate routines in the code that calculate the required timer values based on the requested time delay for period T_{osc} and the necessary offset adjustment. For timer0 (for example) the values are stored in variables txonh and txonl. TMR0H is not the actual high byte of timer0 in 16-bit mode. It is a buffered version of the real high byte of timer0 which is not directly readable or writeable. A write to the high byte of timer0 must take place through the TMR0H buffer register, and the high byte is updated with the contents of TMR0H when a write occurs to TMR0L. This allows all 16 bits of timer0 to be updated at once. Both timers are configured for 16-bit mode, so the same applies to timer1.

Synchronization of the voltage converter takes place by toggling ps_sync in alternate states. The debug port is also controlled in the interrupt routine such that it tracks the TX period.

Before exiting each state, it is necessary to clear the requisite overflow interrupt bit, and to set the next required state.

As you can see from the code below, the PI and DISC modes are interleaved when detect_mode is set to "hybrid". If set to "pulse", the DISC waveform is not sampled, and the detector reverts to a standard PI mode. The variable "hybrid_mode", toggles between

0 and 1 to alternately sample the PI and DISC waveforms. Depending on the state of hybrid_mode, the state machine (when it exits state 0) either moves to state 1 (PI cycle), or state 6 (DISC cycle).

State 0: TX pulse on ... TX pulse width

State 1: TX pulse off ... main sample delay.

State 2: Main sample pulse on ... main sample pulse width.

State 3: Main sample pulse off ... EFE sample delay.

State 4: EFE sample pulse on ... EFE sample pulse width.

State 5: EFE sample pulse off ... wait until timer1 overflows (remainder of TX period).

State 6: TX pulse off ... DISC sample delay.

State 7: DISC sample pulse on ... DISC sample pulse width.

State 8: DISC sample pulse off ... wait until timer1 overflows (remainder of TX period).

```
// Interrupt routine
void interrupt() {
  switch (int_state) {
    case 0:                              // TX on - charge coil
      EN1 = EN1_on;                      // Turn on MOSFET
      TMR0H = txonh;                     // Load TMR0 for TX pulse
      TMR0L = txonl;
      TMR1H = txpdh;                     // Load TMR1 for TX period
      TMR1L = txpdl;
      ps_sync = 1;                       // Sync power supply
      debug = 1;
      PIR1 = 0x00;                       // Clear TMR1 overflow interrupt flag
      INTCON = 0xE0;                     // Clear TMR0 overflow interrupt flag
      if (hybrid_cycle == 0) {
        int_state = 1;                   // Pulse sample
      } else {
        int_state = 6;                   // Disc sample
      }
      break;
    case 1:                              // TX off - discharge coil
      EN1 = EN1_off;                     // Turn off MOSFET
      TMR0H = main_dlyh;                 // Load TMR0 for main sample delay
      TMR0L = main_dlyl;
      ps_sync = 0;                       // Sync power supply
      debug = 0;
      INTCON = 0xE0;                     // Clear TMR0 overflow interrupt flag
      int_state = 2;
      break;
    case 2:                              // Main sample pulse on
      main_pulse = main_on;              // Turn on main sample
      TMR0H = main_smplh;                // Load TMR0 for main sample width
      TMR0L = main_smpll;
      ps_sync = 1;                       // Sync power supply
      INTCON = 0xE0;                     // Clear TMR0 overflow interrupt flag
      int_state = 3;
```

```
      break;
    case 3:                              // Main sample pulse off
      main_pulse = main_off;             // Turn off main sample
      TMR0H = efe_dlyh;                  // Load TMR0 for EFE sample delay
      TMR0L = efe_dlyl;
      ps_sync = 0;                       // Sync power supply
      INTCON = 0xE0;                     // Clear TMR0 overflow interrupt flag
      int_state = 4;
      break;
    case 4:                              // EFE sample pulse on
      efe_pulse = efe_on;                // Turn on EFE sample
      TMR0H = efe_smplh;                 // Load TMR0 for EFE sample width
      TMR0L = efe_smpll;
      ps_sync = 1;                       // Sync power supply
      INTCON = 0xE0;                     // Clear TMR0 overflow interrupt flag
      int_state = 5;
      break;
    case 5:                              // EFE sample pulse off
      efe_pulse = efe_off;               // Turn off EFE sample
      TMR0H = 0x00;                      // Load TMR0 with maximum delay
      TMR0L = 0x00;
      ps_sync = 0;
      INTCON = 0xE0;                     // Clear TMR0 overflow interrupt flag
      if (detect_mode == pulse) {        // Check detector mode (pulse or hybrid)
        hybrid_cycle = 0;                // Continue with pulse sampling
      } else {
        hybrid_cycle = 1;                // Switch to disc sampling
      }
      int_state = 0;
      break;
    case 6:                              // TX off - discharge coil
      EN1 = EN1_off;                     // Turn off MOSFET
      TMR0H = disc_dlyh;                 // Load TMR0 for DISC sample delay
      TMR0L = disc_dlyl;
      ps_sync = 0;                       // Sync power supply
      debug = 0;
      INTCON = 0xE0;                     // Clear TMR0 overflow interrupt flag
      int_state = 7;
      break;
    case 7:                              // Disc sample pulse on
      disc_pulse = disc_on;              // Turn on disc sample pulse
      TMR0H = disc_smplh;                // Load TMR0 for disc sample pulse width
      TMR0L = disc_smpll;
      ps_sync = 1;                       // Sync power supply
      INTCON = 0xE0;                     // Clear TMR0 overflow interrupt flag
      int_state = 8;
      break;
    case 8:                              // Disc sample pulse off
      disc_pulse = disc_off;             // Turn off disc sample pulse
      TMR0H = 0x00;                      // Load TMR0 with maximum delay
      TMR0L = 0x00;
      ps_sync = 0;                       // Sync power supply
      INTCON = 0xE0;                     // Clear TMR0 overflow interrupt flag
      hybrid_cycle = 0;                  // Switch to pulse sampling
      int_state = 0;
      break;
  }
}
```

This is a general display routine that extracts each digit from the value parameter and displays it at the selected row and column on the LCD. It uses the div and mod arithmetic operators available in mikroC PRO to extract the requested digit, and the library function Lcd_Chr is used to display it at the correct location.

```
// Extract requested digit from value and display at selected row and column on LCD
void extract_and_disp(
  unsigned value,                           // Number to display on LCD
  unsigned divisor,                         // Selects which digit to extract
  char row,                                 // Select row on LCD
  char col) {                               // Select column on LCD
  char ch;                                  // Digit to display
  ch = value / divisor % 10;                // Extract selected digit
  Lcd_Chr(row, col, 48 + ch);               // Display selected digit
}
```

As explained in Chapter 3, the PIC can measure voltages up to 5V, and a resistor network is used to divide the battery voltage by 3, so that 15V at the battery pack would register 5V at the ADC. The voltage is a simple average of 8 consecutive readings. In this version the +5V supply rail is used as a reference voltage by the PIC, which means that the voltage reading is not incredibly accurate, but should be good enough to judge whether the battery pack is becoming exhausted. To improve the accuracy, a REF195 bandgap could be used as a reference for the ADC measurements.

```
// Acquire battery voltage
void battery_measure() {
  t = 0;                                    // Clear battery voltage accumulator
  for (i = 1; i <= 8; i++) {
    t += ADC_Read(batt_adc);                // Accumulate battery voltage readings
  }
  t = t / 8;                                // Average readings
  t = (5000 - (t * 5)) * 3 / 100;           // Calculate battery voltage
}
```

The battery voltage is displayed on the LCD using the battery_display routine, which makes use of the extract_and_disp routine discussed earlier.

```
// Show battery voltage on LCD
void battery_display() {
  extract_and_disp(t, 100, 1, 10);
  extract_and_disp(t, 10, 1, 11);
  Lcd_Out(1, 12, ".");
  extract_and_disp(t, 1, 1, 13);
  Lcd_Out(1, 14, "V ");
}
```

In addition to the battery voltage, there is also a battery symbol indicator that shows the battery in various states of charge. If the battery voltage becomes equal to or less than 10.5V, the battery symbol disappears and is replaced by an "X". The battery symbol was created using the custom character generator feature of the LCD. More details of this are provided later.

```
// Display appropriate battery symbol
void battery_symbol() {
  if (t >= 120)
    Lcd_Chr(1, 16, 0);
  else if (t >= 117)
    Lcd_Chr(1, 16, 1);
  else if (t >= 114)
    Lcd_Chr(1, 16, 2);
  else if (t >= 111)
    Lcd_Chr(1, 16, 3);
  else if (t >= 108)
    Lcd_Chr(1, 16, 4);
  else if (t >= 105)
    Lcd_Chr(1, 16, 5);
  else
    Lcd_Out(1, 16, "X");
}
```

Timer0 and timer1 are configured in 16-bit mode, which means that they can hold a maximum value of 65535, or (if you prefer) 2^{16} - 1, or 0xFFFF. When a value is loaded into the registers, the timer starts to count up. When it reaches 65535 an overflow interrupt is generated that diverts program flow to the interrupt routine. If we take calc_txon as an example, this routine calculates the value required to provide the TX mosfet on-time. The variable txon contains the current value in micro-seconds (μs). First the offset is subtracted from the current value, and then converted to clock cycles. Since the PIC is using a 20MHz crystal, the period is 50ns. The program clock cycle is 4 times the period of the crystal, which means we have a T_{osc} of 200ns. For a txon of 150μs and an associated offset of 4μs, timer0 needs to be loaded with:

$$65535 - ((150 - 4) / 2) * 10 = 64805$$

which is 0xFD25 in hexadecimal. In this case, txonh = 0xFD, and txonl = 0x25.

The other calculations follow the same procedure.

```
void calc_txon() {                        // Calculate TX pulse width
  tmp = 65535 - ((txon - txon_offset) / 2) * 10;
  txonh = tmp >> 8;
  txonl = tmp - (txonh << 8);
}
void calc_txpd() {                        // Calculate TX period
  tmp = 65535 - ((txpd - txpd_offset) / 2) * 10;
  txpdh = tmp >> 8;
  txpdl = tmp - (txpdh << 8);
}
```

```
void calc_main_dly() {                          // Calculate main sample delay
  tmp = 65535 - ((main_dly - main_dly_offset) / 2) * 10;
  main_dlyh = tmp >> 8;
  main_dlyl = tmp - (main_dlyh << 8);
}

void calc_main_smpl() {                          // Calculate main sample pulse width
  tmp = 65535 - ((main_smpl - main_smpl_offset) / 2) * 10;
  main_smplh = tmp >> 8;
  main_smpll = tmp - (main_smplh << 8);
}

void calc_efe_dly() {                            // Calculate EFE sample delay
  tmp = 65535 - ((efe_dly - efe_dly_offset) / 2) * 10;
  efe_dlyh = tmp >> 8;
  efe_dlyl = tmp - (efe_dlyh << 8);
}

void calc_efe_smpl() {                           // Calculate EFE sample pulse width
  efe_smplh = main_smplh;                        // EFE sample pulse width must equal main sample
  efe_smpll = main_smpll;
}

void calc_disc_dly() {                           // Calculate disc sample delay
  tmp = 65535 - ((disc_dly - disc_dly_offset) / 2) * 10;
  disc_dlyh = tmp >> 8;
  disc_dlyl = tmp - (disc_dlyh << 8);
}

void calc_disc_smpl() {                          // Calculate disc sample pulse width
  tmp = 65535 - ((disc_smpl - disc_smpl_offset) / 2) * 10;
  disc_smplh = tmp >> 8;
  disc_smpll = tmp - (disc_smplh << 8);
}
```

The PI and DISC readings are stored in arrays to provide a running average. This routine clears the arrays, pointers, and array total at the start of the program, and also when leaving the menu system.

```
void clear_arrays() {                            // Clear both PI and disc arrays
  for (i = 0; i <= (pi_num - 1); i++) {
    pi_array[i] = 0;                             // Clear PI array
    disc_array[i] = 0;                           // Clear disc array
  }
  pi_pointer = 0;                                // Reset PI array pointer
  pi_total = 0;                                  // Clear PI array total
  disc_pointer = 0;                              // Reset disc array pointer
  disc_total = 0;                                // Clear disc array total;
}
```

The detector settings are stored in the PIC's internal EEPROM so that it remembers any changes the user has made via the menu system. The first location is set to 0xAA to indicate that the EEPROM has been initialized. If not, at startup it loads the EEPROM with default values.

The Voodoo Project

```
void write_eeprom() {
  EEPROM_Write(0, 0xAA);              // Initialize EEPROM
  EEPROM_Write(1, detect_mode);       // Write detection mode
  tmp = pi_thr >> 8;                  // Extract PI threshold high byte
  EEPROM_Write(2, tmp);               // Write PI threshold high byte
  tmp = (pi_thr << 8) >> 8;           // Extract PI threshold low byte
  EEPROM_Write(3, tmp);               // Write PI threshold low byte
  EEPROM_Write(4, txon);              // Write TX pulse width
  tmp = txpd >> 8;                    // Extract TX period high byte
  EEPROM_Write(5, tmp);               // Write TX period high byte
  tmp = (txpd << 8) >> 8;             // Extract TX period low byte
  EEPROM_Write(6, tmp);               // Write TX period low byte
  EEPROM_Write(7, main_dly);          // Write main sample delay
  EEPROM_Write(8, main_smpl);         // Write main sample pulse width
  tmp = efe_dly >> 8;                 // Extract EFE sample delay high byte
  EEPROM_Write(9, tmp);               // Write EFE sample delay high byte
  tmp = (efe_dly << 8) >> 8;          // Extract EFE sample delay low byte
  EEPROM_Write(10, tmp);              // Write EFE sample delay low byte
  EEPROM_Write(11, pi_num);           // Write number of PI readings to average
  tmp = disc_thr >> 8;                // Extract disc threshold high byte
  EEPROM_Write(12, tmp);              // Write disc threshold high byte
  tmp = (disc_thr << 8) >> 8;         // Extract disc threshold low byte
  EEPROM_Write(13, tmp);              // Write disc threshold low byte
  EEPROM_Write(14, disc_dly);         // Write disc sample delay
  EEPROM_Write(15, disc_smpl);        // Write disc sample pulse width
  EEPROM_Write(16, disc_num);         // Write number of disc readings to average
  EEPROM_Write(17, meter_zero_limit); // Write meter zero limit
  EEPROM_Write(18, counter_limit);    // Write accept and reject counter limit
}
```

Key bounce can occur in mechanical switches and can cause one switch press to be detected as multiple presses. This happens with the push buttons of the menu system, and it is necessary to use "debounce" software to eliminate these false button pushes. The unwanted bounce occurs both when the button is pressed and when it is released, so the software needs to take account of this. A button press is recorded when it is released, and the button number is returned to the calling function. Delay_ms() is a library function that can be used to provide a software delay, defined in milli-seconds.

```
// Scan keypad for a key press
char scan_keypad() {
  result = 0;
  if (menu_btn_port == btn_on) {          // Check menu button
    result = menu_btn;
    Delay_ms(debounce);
    while (menu_btn_port == btn_on) {}
    Delay_ms(debounce);
  } else {
    if (up_btn_port == btn_on) {          // Check up button
      result = up_btn;
      Delay_ms(debounce);
      while (up_btn_port == btn_on) {}
      Delay_ms(debounce);
    } else {
      if (down_btn_port == btn_on) {      // Check down button
        result = down_btn;
        Delay_ms(debounce);
        while (down_btn_port == btn_on) {}
        Delay_ms(debounce);
      } else {
        if (enter_btn_port == btn_on) {   // Check enter button
```

```
            result = enter_btn;
            Delay_ms(debounce);
            while (enter_btn_port == btn_on) {}
            Delay_ms(debounce);
        }
      }
    }
  }
  return result;
}
```

The menu system allows the user to switch the detecting mode between PI and HYBRID, and to change a number of parameters. Also note that the timer interrupts are disabled upon entering the menu system, the mosfet is turned off to save power, and the audio is disabled.

The menu system is entered by pressing and releasing the Menu button, and likewise to exit. The Up button increases the value of any parameter, and the Down button does the opposite. The Enter button moves the menu to the next parameter. If the detector is in PI mode only the parameters relevant to that mode are displayed. In Hybrid mode only parameters related to discrimination are shown.

Upon exiting the menu system the EEPROM is updated with the latest settings, the timer interrupts are enabled, and the interrupt state machine is reset to 0.

PI Threshold – When the input voltage is read, it is returned as a number between 0 and 1023 (as 2^{10} = 1024). Since the maximum input voltage is 5V, the ADC resolution is 5V/1024 = 4.88mV. The output of the PI channel swings between +5V and -5V, with its no target voltage hovering around 0V. A resistor divider network converts this voltage so that it swings between +5V and 0V (referenced to the -5V supply rail) with the no target voltage hovering around +2.5V. In this case the lower end of the PI threshold range is set to 511, because 511 * 4.88mV = 2.49V. The upper limit to the range is set to 531, which equates to 531 * 4.88mV = 2.59V. The default value is 525. To make these numbers more meaningful to the user, only the difference between the PI threshold and 511 is displayed on the LCD so that the threshold range will appear to be 0 to 20.

PI Pulse Width - The TX pulse width can be adjusted from 10μs to 180μs. The default value is 150μs.

PI Sample Delay – The time between mosfet switch-off and the start of the main sample pulse can be adjusted from 15μs to 50μs. The default value is 27μs to match the Troy coil.

PI Sample Width – The main sample width has an effect on the gain of the signal, and can be adjusted from 20μs to 60μs. The default value is 60μs.

PI Average – This defines the size of the running average array. It can be adjusted from 1 to 64. The default value is 64.

DISC Threshold – The DISC channel threshold settings are identical to the PI threshold, and assumes the amplitude of the DISC channel increases in a positive direction for a non-ferrous target.

DISC Sample Delay – A delay is required between mosfet switch-off and the start of the second half-cycle of the decaying sine wave. The range of adjustment is from 20μs to 100μs, and the default setting is 45μs.

DISC Sample Width – The sample width needs to encompass the full half-cycle. Range of adjustment is from 10μs to 100μs, to allow for flexibility during experimentation. The default value is 45μs.

DISC Average – Like the PI average, this defines the size of the running average array. It can be adjusted from 1 to 64. The default value is 64.

Meter Zero Limit – After a period of time when no targets have been detected, the LCD meter starts to automatically zero by removing blocks from the display. The meter zero limit defines how many non-target readings need to occur before auto-zeroing starts to take place. The range of adjustment is between 5 and 10. The default value is 10.

Counter Limit – This is the number of targets detected before a block is added to the LCD meter, either accept (non-ferrous) or reject (ferrous). The range is 10 to 150, and the default setting is 100. This value affects the responsiveness of the meter.

```
// Menu system
void menu_system() {
  PIE1 = 0x00;                          // Disable TMR1 interrupt
  INTCON = 0x00;                        // Disable global, peripheral and TMR0 interrupts
  menu_flag = menu_active;              // Flag menu system as active
  EN1 = EN1_off;                        // Turn off MOSFET to save power
  audio_en = audio_off;                 // Disable audio while in menu system
  menu_disp = 0;                        // Set menu display to first screen
  while (menu_flag == menu_active) {    // Enter menu system
    switch (menu_disp) {
    case 0:
      Lcd_Out(1, 1, " DETECTING MODE ");
      if (detect_mode == pulse) {
        Lcd_Out(2, 1, "     PULSE      ");
      } else {
        Lcd_Out(2, 1, "     HYBRID     ");
      }
      if ((keypad_btn == up_btn) || (keypad_btn == down_btn)) { // Toggle detection mode
```

```
        if (detect_mode == pulse) {
          detect_mode = hybrid;
        } else {
          detect_mode = pulse;
        }
      }
      break;
    case 1:
      Lcd_Out(1, 1, "   PI THRESHOLD   ");
      Lcd_Out(2, 1, "            ");
      thr_disp = pi_thr - 511;
      extract_and_disp(thr_disp, 10, 2, 8);
      extract_and_disp(thr_disp, 1, 2, 9);
      Lcd_Out(2, 10, "          ");
      if (keypad_btn == up_btn) {
        pi_thr++;
        if (pi_thr > 531) {pi_thr = 531;}
      }
      if (keypad_btn == down_btn) {
        pi_thr--;
        if (pi_thr < 511) {pi_thr = 511;}
      }
      break;
    case 2:
      Lcd_Out(1, 1, " PI PULSE WIDTH ");
      Lcd_Out(2, 1, "          ");
      extract_and_disp(txon, 100, 2, 6);
      extract_and_disp(txon, 10, 2, 7);
      extract_and_disp(txon, 1, 2, 8);
      Lcd_Out(2, 9, "us        ");
      if (keypad_btn == up_btn) {
        txon += 10;
        if (txon > 180) {txon = 180;}
      }
      if (keypad_btn == down_btn) {
        txon -= 10;
        if (txon < 10) {txon = 10;}
      }
      break;
    case 3:
      Lcd_Out(1, 1, "PI SAMPLE DELAY ");
      Lcd_Out(2, 1, "        ");
      extract_and_disp(main_dly, 10, 2, 7);
      extract_and_disp(main_dly, 1, 2, 8);
      Lcd_Out(2, 9, "us        ");
      if (keypad_btn == up_btn) {
        main_dly++;
        if (main_dly > 50) {main_dly = 50;}
      }
      if (keypad_btn == down_btn) {
        main_dly--;
        if (main_dly < 15) {main_dly = 15;}
      }
      break;
    case 4:
      Lcd_Out(1, 1, "PI SAMPLE WIDTH ");
      Lcd_Out(2, 1, "       ");
      extract_and_disp(main_smpl, 100, 2, 6);
      extract_and_disp(main_smpl, 10, 2, 7);
      extract_and_disp(main_smpl, 1, 2, 8);
      Lcd_Out(2, 9, "us        ");
      if (keypad_btn == up_btn) {
        main_smpl++;
        if (main_smpl > 60) {main_smpl = 60;}
      }
      if (keypad_btn == down_btn) {
        main_smpl--;
        if (main_smpl < 20) {main_smpl = 20;}
      }
      break;
    case 5:
      Lcd_Out(1, 1, "   PI AVERAGE   ");
```

```
  Lcd_Out(2, 1, "        ");
  extract_and_disp(pi_num, 10, 2, 8);
  extract_and_disp(pi_num, 1, 2, 9);
  Lcd_Out(2, 10, "        ");
  if (keypad_btn == up_btn) {
    pi_num++;
    if (pi_num > 64) {pi_num = 64;}
  }
  if (keypad_btn == down_btn) {
    pi_num--;
    if (pi_num < 1) {pi_num = 1;}
  }
  break;
case 6:
  Lcd_Out(1, 1, " DISC THRESHOLD ");
  Lcd_Out(2, 1, "        ");
  thr_disp = disc_thr - 511;
  extract_and_disp(thr_disp, 10, 2, 8);
  extract_and_disp(thr_disp, 1, 2, 9);
  Lcd_Out(2, 10, "        ");
  if (keypad_btn == up_btn) {
    disc_thr++;
    if (disc_thr > 531) {disc_thr = 531;}
  }
  if (keypad_btn == down_btn) {
    disc_thr--;
    if (disc_thr < 511) {disc_thr = 511;}
  }
  break;
case 7:
  Lcd_Out(1, 1, "DISC SMPL DELAY ");
  Lcd_Out(2, 1, "        ");
  extract_and_disp(disc_dly, 100, 2, 6);
  extract_and_disp(disc_dly, 10, 2, 7);
  extract_and_disp(disc_dly, 1, 2, 8);
  Lcd_Out(2, 9, "us        ");
  if (keypad_btn == up_btn) {
    disc_dly++;
    if (disc_dly > 100) {disc_dly = 100;}
  }
  if (keypad_btn == down_btn) {
    disc_dly--;
    if (disc_dly < 20) {disc_dly = 20;}
  }
  break;
case 8:
  Lcd_Out(1, 1, "DISC SMPL WIDTH ");
  Lcd_Out(2, 1, "        ");
  extract_and_disp(disc_smpl, 100, 2, 6);
  extract_and_disp(disc_smpl, 10, 2, 7);
  extract_and_disp(disc_smpl, 1, 2, 8);
  Lcd_Out(2, 9, "us        ");
  if (keypad_btn == up_btn) {
    disc_smpl++;
    if (disc_smpl > 100) {main_smpl = 100;}
  }
  if (keypad_btn == down_btn) {
    disc_smpl--;
    if (disc_smpl < 10) {disc_smpl = 10;}
  }
  break;
case 9:
  Lcd_Out(1, 1, "  DISC AVERAGE  ");
  Lcd_Out(2, 1, "        ");
  extract_and_disp(disc_num, 10, 2, 8);
  extract_and_disp(disc_num, 1, 2, 9);
  Lcd_Out(2, 10, "        ");
  if (keypad_btn == up_btn) {
    disc_num++;
    if (disc_num > 64) {disc_num = 64;}
  }
  if (keypad_btn == down_btn) {
```

```
          disc_num--;
          if (disc_num < 1) {disc_num = 1;}
        }
        break;
      case 10:
        Lcd_Out(1, 1, "METER ZERO LIMIT");
        Lcd_Out(2, 1, "          ");
        extract_and_disp(meter_zero_limit, 10, 2, 8);
        extract_and_disp(meter_zero_limit, 1, 2, 9);
        Lcd_Out(2, 10, "        ");
        if (keypad_btn == up_btn) {
          meter_zero_limit++;
          if (meter_zero_limit > 20) {meter_zero_limit = 20;}
        }
        if (keypad_btn == down_btn) {
          meter_zero_limit--;
          if (meter_zero_limit < 5) {meter_zero_limit = 5;}
        }
        break;
      case 11:
        Lcd_Out(1, 1, " COUNTER LIMIT  ");
        Lcd_Out(2, 1, "        ");
        extract_and_disp(counter_limit, 100, 2, 7);
        extract_and_disp(counter_limit, 10, 2, 8);
        extract_and_disp(counter_limit, 1, 2, 9);
        Lcd_Out(2, 10, "        ");
        if (keypad_btn == up_btn) {
          counter_limit += 10;
          if (counter_limit > 150) {counter_limit = 150;}
        }
        if (keypad_btn == down_btn) {
          counter_limit -= 10;
          if (counter_limit < 10) {counter_limit = 10;}
        }
        break;
    }
    keypad_btn = scan_keypad();                  // Scan the keypad
    if (keypad_btn == menu_btn) {
      menu_flag = menu_inactive;                 // Deactivate menu system if menu button pressed
      write_eeprom();                            // Write data to EEPROM
      calc_txon();                               // Calculate TX pulse width and set timer
      calc_txpd();                               // Calculate TX period and set timer
      calc_main_dly();                           // Calculate main sample delay and set timer
      calc_main_smpl();                          // Calculate main sample pulse width and set timer
      calc_efe_dly();                            // Calculate EFE sample delay and set timer
      calc_efe_smpl();                           // Calculate EFE sample pulse width and set timer
      calc_disc_dly();                           // Calculate disc sample delay and set timer
      calc_disc_smpl();                          // Calculate disc sample pulse width and set timer
      clear_arrays();                            // Clear both PI and DISC arrays
      int_state = 0;                             // Reset interrupt state machine
      PIE1 = 0x01;                               // Enable TMR0 interrupt
      INTCON = 0xE0;                             // Enable global, peripheral and TMR1 interrupts
    } else {
      if (keypad_btn == enter_btn) {             // Navigate menu system
        switch (menu_disp) {
          case 0:                                // Detect mode
            if (detect_mode == hybrid) {
              menu_disp = 6;                      // Go to hybrid settings
            } else {
              menu_disp = 1;                      // Go to pulse settings
            }
            break;
          case 1:                                // PI threshold
            menu_disp = 2;
            break;
          case 2:                                // PI pulse width
            menu_disp = 3;
            break;
          case 3:                                // PI sample delay
            menu_disp = 4;
            break;
          case 4:                                // PI sample pulse width
```

```
          menu_disp = 5;
          break;
        case 5:                          // PI running average
          menu_disp = 0;
          break;
        case 6:                          // Disc threshold
          menu_disp = 7;
          break;
        case 7:                          // Disc sample delay
          menu_disp = 8;
          break;
        case 8:                          // Disc sample pulse width
          menu_disp = 9;
          break;
        case 9:
          menu_disp = 10;                // Disc running average
          break;
        case 10:                         // Meter zero limit
          menu_disp = 11;
          break;
        case 11:                         // Accept and reject counter limit
          menu_disp = 0;
          break;
        }
      }
    }
  }
}
```

This routine resets the LCD by displaying either PULSE or HYBRID on the top line, to represent the current mode of operation, and the bottom line is cleared, except for the middle marker (vertical line) of the meter.

```
void disp_reset() {
  if (detect_mode == pulse) {
    Lcd_Out(1, 1, " PULSE    ");
  } else {
    Lcd_Out(1, 1, " HYBRID  ");
  }
  Lcd_Out(2, 1, "          ");
  Lcd_Chr(2, 8, 7);                      // Display middle marker (vertical line)
  Lcd_Out(2, 9, "          ");
}
```

Finally we reach the main section of code.

The first job is to initialize the ports by setting them to either inputs or outputs as required.

```
// Main program section
void main() {
  // Initialize ports
  STRIA = TRISA_INIT;
  TRISB = TRISB_INIT;
  TRISC = TRISC_INIT;
  TRISD = TRISD_INIT;
```

```
TRISE = TRISE_INIT;
```

Next the mosfet, and all the sample pulses are turned off.

```
// Turn off MOSFET and sample pulses
EN1 = EN1_off;                          // Turn off MOSFET
main_pulse = main_off;                  // Turn off main sample pulse
efe_pulse = efe_off;                    // Turn off EFE sample pulse
disc_pulse = disc_off;                  // Turn off disc sample pulse
```

The ADCON register is then set to enable AD0 to AD7, although we are only using AD0, AD1 and AD7. There is no option to disable the unused ADCs.

```
// Initialize ADCs
ADCON1 = 0x07;                          // Enable AD0 to AD7 (VDD and VSS as voltage ref)
ADC_Init();
```

The library provides a number of functions to control the LCD.

Lcd_Init() initializes the LCD to use the 4-wire interface according to the variables defined earlier. Two commands are then sent to the LCD to clear the display and to turn off the flashing cursor.

Although the LCD has a built-in default font, there are some symbols missing that we require for this project. Symbols can be loaded into the LCD's character generator (CG) RAM. Mikroelektronika's mikroC PRO includes a useful LCD Custom Char Generator that can assist in creating the necessary characters via a graphical interface. There is a limit of 8 custom characters that can be loaded into CGRAM.

The characters we require are various battery symbols to indicate the state of the battery voltage, a block character for the LCD meter, and a vertical line for the centre line of the display.

```
// Initialize LCD
Lcd_Init();
Lcd_Cmd(_LCD_CLEAR);
Lcd_Cmd(_LCD_CURSOR_OFF);

// Load custom characters into CG RAM of LCD
Lcd_Cmd(64);
// Battery >= 12.0V (character 0)
Lcd_Chr_Cp(14);
```

```
for (i = 2; i <= 8; i++) {Lcd_Chr_Cp(31);}
// Battery >= 11.7V (character 1)
Lcd_Chr_Cp(14);
Lcd_Chr_Cp(31);
Lcd_Chr_Cp(17);
for (i = 4; i <= 8; i++) {Lcd_Chr_Cp(31);}
// Battery >= 11.4V (character 2)
Lcd_Chr_Cp(14);
Lcd_Chr_Cp(31);
for (i = 3; i <= 4; i++) {Lcd_Chr_Cp(17);}
for (i = 5; i <= 8; i++) {Lcd_Chr_Cp(31);}
// Battery >= 11.1V (character 3)
Lcd_Chr_Cp(14);
Lcd_Chr_Cp(31);
for (i = 3; i <= 5; i++) {Lcd_Chr_Cp(17);}
for (i = 6; i <= 8; i++) {Lcd_Chr_Cp(31);}
// Battery >= 10.8V (character 4)
Lcd_Chr_Cp(14);
Lcd_Chr_Cp(31);
for (i = 3; i <= 6; i++) {Lcd_Chr_Cp(17);}
for (i = 7; i <= 8; i++) {Lcd_Chr_Cp(31);}
// Battery >= 10.5V (character 5)
Lcd_Chr_Cp(14);
Lcd_Chr_Cp(31);
for (i = 3; i <= 7; i++) {Lcd_Chr_Cp(17);}
Lcd_Chr_Cp(31);
// Block symbol for ferrous /non-ferrous display (character 6)
for (i = 1; i <= 8; i++) {Lcd_Chr_Cp(31);}
// Middle character on ferrous / non-ferrous display (character 7)
for (i = 1; i <= 8; i++) {Lcd_Chr_Cp(4);}
```

When the detector is switched on, a splash screen is shown for 2 seconds, displaying the current software version (e.g. "VOODOO V1.0") on the top line, and "HYBRID DETECTOR" on the second line.

```
// Display splash screen
Lcd_Out(1,1,"  VOODOO V1.0   ");
Lcd_Out(2,1,"HYBRID DETECTOR ");
delay_ms(2000);
Lcd_Cmd(_LCD_CLEAR);
```

A number of variables are then initialized (reset) to zero.

```
// Initialize variables
int_state = 0;          // Initialize interrupt state machine
main_state = 0;         // Initialize main state machine
loop_count = 0;         // Reset loop counter
hybrid_cycle = 0;       // Hybrid operating cycle (0 = pulse, 1 = disc)
accept = 0;             // Clear accept (non-ferrous) counter
accept_blk = 0;         // Clear accept block display counter
reject = 0;             // Clear reject (ferrous) counter
reject_blk = 0;         // Clear reject block display counter
update_disp = 0;        // Clear update display flag
meter_zero = 0;         // Clear meter zero counter
```

The various detector parameters are then initialized with default values, but these may be overwritten later if they differ from the values stored in EEPROM.

```
// All default settings below may be overwritten by EEPROM
detect_mode = hybrid;            // Set default detection mode to hybrid
pi_thr = 525;                    // Set default pi threshold
disc_thr = 511;                  // Set default disc threshold
txon = 150;                      // Set default TX pulse width (us)
txpd = 1000;                     // Set default TX pulse period (us)
main_dly = 27;                   // Set default main sample delay (us)
main_smpl = 60;                  // Set default main sample pulse width (us)
efe_dly = 650;                   // Set default EFE sample delay (us)
pi_num = 64;                     // Set default number of PI readings to average
disc_dly = 45;                   // Set default disc sample delay (us)
disc_smpl = 45;                  // Set default disc sample pulse width (us)
disc_num = 64;                   // Set default number of disc readings to average
meter_zero_limit = 10;           // Set meter zero limit
counter_limit = 100;             // Set accept and reject counter limit
```

Both the PI and DISC running average arrays need to be cleared, otherwise the calculated values will be incorrect.

```
clear_arrays();                  // Clear both PI and DISC arrays
```

Now the saved parameters must be loaded from EEPROM. This is only done if the first EEPROM location contains 0xAA to show that it has been initialized. If not, then this is the first time the detector has been switched on following a software install, and the default values are stored in EEPROM and used instead.

```
// Load saved values from EEPROM
if (EEPROM_Read(0) == 0xAA) {        // Check if EEPROM already initialized
    detect_mode = EEPROM_Read(1);    // Read detection mode (pulse or hybrid)
    pi_thr = EEPROM_Read(2);         // Read PI threshold high byte
    pi_thr = pi_thr << 8;
    pi_thr += EEPROM_Read(3);        // Read PI threshold low byte
    txon = EEPROM_Read(4);           // Read TX pulse width
    txpd = EEPROM_Read(5);           // Read TX period high byte
    txpd = txpd << 8;
    txpd += EEPROM_Read(6);          // Read TX period low byte
    main_dly = EEPROM_Read(7);       // Read main sample delay
    main_smpl = EEPROM_Read(8);      // Read main sample pulse width
    efe_dly = EEPROM_Read(9);        // Read EFE sample delay high byte
    efe_dly = efe_dly << 8;
    efe_dly += EEPROM_Read(10);      // Read EFE sample delay low byte
    pi_num = EEPROM_Read(11);        // Read number of PI readings to average
    disc_thr = EEPROM_Read(12);      // Read disc threshold high byte
    disc_thr = disc_thr << 8;
    disc_thr += EEPROM_Read(13);     // Read disc threshold low byte
    disc_dly = EEPROM_Read(14);      // Read disc sample delay
    disc_smpl = EEPROM_Read(15);     // Read disc sample pulse width
    disc_num = EEPROM_Read(16);      // Read number of disc readings to average
    meter_zero_limit = EEPROM_Read(17);  // Read meter zero limit
    counter_limit = EEPROM_Read(18); // Read accept and reject counter limit
} else{
```

```
    write_eeprom();                           // Write settings to EEPROM
}
```

Following successful loading of the detector parameters from EEPROM, the timer values are calculated and the timers are set. Finally, the LCD is reset ready for operation.

```
calc_txon();                  // Calculate TX pulse width and set timer
calc_txpd();                  // Calculate TX period and set timer
calc_main_dly();              // Calculate main sample delay and set timer
calc_main_smpl();             // Calculate main sample pulse width and set timer
calc_efe_dly();               // Calculate EFE sample delay and set timer
calc_efe_smpl();              // Calculate EFE sample pulse width and set timer
calc_disc_dly();              // Calculate disc sample delay and set timer
calc_disc_smpl();             // Calculate disc sample pulse width and set timer
disp_reset();                 // Reset display
```

The timers are initialized next, the interrupts enabled, and the interrupt overflow bits cleared.

```
// Initialize timers and interrupts
// TMR0 used to generate sample pulses and delays (main and EFE) and TX pulses
// TMR1 used to generate TX period
T0CON = 0x88;                 // Configure TMR0 as 16-bit
T1CON = 0x85;                 // Configure TMR1 as 16-bit
TMR0H = txonh;                // Load TX pulse width
TMR0L = txonl;
TMR1H = txpdh;                // Load TX period
TMR1H = txpdl;
PIR1 = 0x00;                  // Clear TMR1 interrupt overflow flag
PIE1 = 0x01;                  // Enable TMR1 overflow interrupt
INTCON = 0xE0;                // Enable interrupts
```

At long last we get to the infinite while loop that represents the main program. This section of code is also controlled by a state machine, but with only two states.

At the start of state 0 we first acquire the PI target reading and save it in the PI running average array. The running average algorithm uses a rotating pointer to update the next value, and to calculate the array total. In that way you don't have to total up the array each time, and it's much faster.

If the detector is running in hybrid mode, the same procedure is performed for the DISC readings.

```
while(1) {
  switch (main_state) {
```

Embedded Software   **63**

```
case 0:
    pi_target = Adc_Read(pi_adc);              // Acquire PI target reading
    pi_tmp = pi_array[pi_pointer];             // Save current PI array value
    pi_array[pi_pointer] = pi_target;          // Add latest reading to PI array
    pi_total = (pi_total + pi_target) - pi_tmp; // Calculate PI array total
    pi_target = pi_total / pi_num;             // Average readings
    pi_pointer++;                              // Increment PI pointer
    if (pi_pointer >= pi_num) {
      pi_pointer = 0;                          // Reset PI array pointer
    }
    if (detect_mode == hybrid) {
      disc_target = Adc_Read(disc_adc);        // Acquire DISC target reading
      disc_tmp = disc_array[disc_pointer];     // Save current disc array value
      disc_array[disc_pointer] = disc_target;  // Add latest reading to disc array
      disc_total = (disc_total + disc_target) - disc_tmp; // Calculate disc array total;
      disc_target = disc_total / disc_num;     // Average readings
      disc_pointer++;                          // Increment disc pointer
      if (disc_pointer >= disc_num) {
        disc_pointer = 0;                      // Reset disc array pointer
      }
    }
```

The current PI average (pi_target) is compared with the PI threshold, and also the current DISC average (disc_target) is compared with the DISC threshold if the detector is in hybrid mode. If both the PI and DISC averages are above their respective thresholds, the audio output is enabled, and the LCD meter is updated by either increasing the number of accept blocks or decreasing the number of reject blocks with a dependency on the counter limit. A similar process occurs if the target is determined to be ferrous, with an increase in reject blocks, or a decrease in accept blocks whichever is appropriate. The LCD display is only updated if the number of accept or reject blocks changes. This is controlled by the variable update_disp.

The comments in the code should help you to unravel how it works.

```
if (pi_target >= pi_thr) {                     // Decide whether to beep or not
    if (detect_mode == hybrid) {               // Check for hybrid mode
      update_disp = 0;                         // Reset update display flag
      if (disc_target >= disc_thr) {           // Non-ferrous target detected
        audio_en = audio_on;                   // Beep
        if (reject_blk == 0) {
          accept++;                            // Increment accept counter if no reject blocks
          reject = 0;                          // Clear reject counter
          if (accept > counter_limit) {        // Check if accept counter has reached limit
            accept = 0;                        // Reset accept counter to zero
            accept_blk++;                      // Increment number of accept blocks
            if (accept_blk <= blk_limit) {
              update_disp = 1;                 // Set update display flag
            } else {
              accept_blk = blk_limit;          // Limit number of accept blocks
            }
          }
        } else {                               // There must be some reject blocks displayed
          if (reject != 0) {
            reject--;                          // Decrement reject counter if not at zero
          } else {                             // Reject counter must have reached zero
            reject = counter_limit;            // Set reject counter to limit
            if (reject_blk != 0) {
```

```
            reject_blk--;                  // Decrement number of reject blocks if not zero
            update_disp = 1;               // Set update display flag
          }
        }
      }
    } else {                               // Ferrous target detected
      audio_en = audio_off;                // Do not beep
      if (accept_blk == 0) {
        reject++;                          // Increment reject counter if no accept blocks
        accept = 0;                        // Clear accept counter
        if (reject > counter_limit) {      // Check if reject counter has reached limit
          reject = 0;                      // Reset reject counter to zero
          reject_blk++;                    // Increment number of reject blocks
          if (reject_blk <= blk_limit) {
            update_disp = 1;               // Set update display flag
          } else {
            reject_blk = blk_limit;        // Limit number of reject blocks
          }
        }
      } else {                             // There must be some accept blocks displayed
        if (accept != 0) {
          accept--;                        // Decrement accept counter is not at zero
        } else {                           // Accept counter must have reached zero
          accept = counter_limit;          // Set accept counter to limit
          if (accept_blk != 0) {
            accept_blk--;                  // Decrement number of accept blocks if not zero
            update_disp = 1;               // Set update display flag
          }
        }
      }
    }
  } else {                                 // Mode must be PI
    audio_en = audio_on;                   // Beep
  }
} else {                                   // No target detected
  audio_en = audio_off;                    // Do not beep
}
if (update_disp == 0) {                    // Only zero display when no signal is present
  meter_zero++;                            // Increment meter zero counter
  if (meter_zero >= meter_zero_limit) {
    if (accept_blk != 0) {                 // Must be accept blocks displayed
      if (accept != 0) {
        accept--;                          // Decrement accept counter if not already zero
      } else {
        accept_blk--;                      // Decrement number of accept blocks
        accept = counter_limit;            // Set accept counter to limit
        update_disp = 1;                   // Set update display flag
      }
    } else {
      if (reject_blk != 0) {               // Otherwise must be reject blocks displayed
        if (reject != 0) {
          reject--;                        // Decrement reject counter if not already zero
        } else {
          reject_blk--;                    // Decrement number of reject blocks displayed
          reject = counter_limit;          // Set reject counter to limit
          update_disp = 1;                 // Set update display flag
        }
      }
    }
    meter_zero = 0;                        // Reset meter zero counter
  }
}
if (update_disp == 1) {                    // Check if display needs to be updated
  if (accept_blk != 0) {
    for (i = 1; i <= accept_blk; i++) {
      Lcd_Chr(2, 8 + i, 6);                // Display required number of non-ferrous blocks
    }
    if (accept_blk != blk_limit) {
      for (i = accept_blk + 1; i <= blk_limit; i++) {
        Lcd_Chr(2, 8 + i, 32);             // Fill rest of non-ferrous display with spaces
      }
```

```
      }
    } else {
      Lcd_Chr(2, 9, 32);                    // Clear first block in non-ferrous display
      if (reject_blk != 0) {
        for (i = 1; i <= reject_blk; i++) {
          Lcd_Chr(2, 8 - i, 6);             // Display required number of ferrous blocks
        }
        if (reject_blk != blk_limit) {
          for (i = reject_blk + 1; i <= blk_limit; i++) {
            Lcd_Chr(2, 8 - i, 32);          // Fill rest of ferrous display with spaces
          }
        }
      } else {
        Lcd_Chr(2, 7, 32);                  // Clear first block in ferrous display
      }
    }
    update_disp = 0;                        // Reset update display flag
  }
```

After acquiring the target signals and updating the LCD meter, the keypad is scanned to check if the menu button has been pressed.

```
keypad_btn = scan_keypad();
if (keypad_btn == menu_btn) {              // Check if menu button pressed
  menu_system();                           // Enter menu system
  disp_reset();                            // Reset display after exiting menu system
}
```

The loop count is incremented, and the main code's state machine moves to state 1 if the count has reached 128.

```
loop_count++;                              // Increment loop count
main_state = (loop_count == 128)?1:0;      // Read battery voltage after loop count of 128
break;
```

In state 1 the battery voltage is measured and displayed, both as a voltage and using the symbols that were custom generated earlier and placed in the LCD's CGRAM.

The main state machine then moves back to state 0.

```
case 1:
  battery_measure();                       // Acquire battery voltage
  battery_display();                       // Display battery voltage
  battery_symbol();                        // Display appropriate battery symbol
  loop_count = 0;                          // Reset loop count
  main_state = 0;                          // Reset main program state machine
  break;
  }
 }
}
```

The Voodoo Project

When mikroC Pro for PIC compiles the source code to hex, this includes the EEPROM data, user bytes (user ID), and the configuration words. This prevents any problems with having to set configuration (fuse) bits during the programming procedure.

For reference, 55% of RAM and 27% of ROM within the PIC18F4520 was used by the Voodoo software.

Chapter 6   *Coil Testing and Construction*

"Show me to the non-ferrous metals, mate!"

--- Lance (2014) – British TV Comedy: *Detectorists*

Fig. 6-1: Troy Concentric Coil

Troy Concentric Coil

As mentioned in previous chapters, a 9" diameter concentric Troy coil (Fig. 6-1) was used for all the initial experiments with the Voodoo detector. Despite having a high TX loop inductance of 1.12mH this was one of the best coils tested in the field, and it was easily able to detect the first two coins in the test garden. This is a good result bearing in mind that no other detector has ever been able to reliably detect the third coin, with the fourth and fifth coins being a virtual impossibility.

The PI channel main sample delay was set to 27µs which makes this quite a slow coil, but perfectly adequate for coin shooting. At the DISC preamp output the first positive peak gave an amplitude decrease for non-ferrous metals, and an increase for ferrous. Note that these are the characteristics we will be seeking when testing various coils in this chapter. Depending on the coil being used, the DISC sampling may either need to occur on the positive half-cycle of the ringing waveform or on the negative half-cycle, although in one case it was necessary to sample on the rising edge.

The ringing frequency of the DISC waveform was 10.5kHz (with a parallel capacitance of 33nF) and the DISC delay was set to 45µs.

The different measurements and settings for the coils tested are shown in Table 6-1 at the end of the chapter.

Fig. 6-2: Fisher F75 Coil

Fisher F75 Coil

The Fisher F75 coil is a DD type (Fig. 6-2), which may not appear to be the ideal choice for a hybrid detector that can be switched into standard PI mode. This is because the detector will effectively be operating with a mono coil that is offset from the centre of the shaft. Despite this potential problem the Fisher coil was found to be very good at locating the first two coins in the test garden, and the offset issue was not that apparent when being used as a standard PI.

The PI channel main sample delay was set to 18μs, which was considerably faster than the Troy coil. This was no doubt due to the fact that the TX (PI) loop had a much lower inductance of 697μH. The inductance of the RX (DISC) loop was 6.9mH, with a parallel capacitance of 33nF resulting in a ringing frequency of 10.5kHz. The DISC sample delay was set to 45μs, and the DISC preamp amplitude on the first negative peak was equal to -440mV.

With the Troy coil I had replaced the connector and wired it to match the requirements of Voodoo. However I did not wish to do this with the Fisher F75 coil, and decided (in this case) to fit an adapter cable. The wiring of this cable is shown in Fig. 6-7.

Fig. 6-3: Elliptical Coil

Elliptical Coil

To quote Dave Johnson (Chief Designer, First Texas Products and Fisher Research Labs):
"For finding small objects and for searching in an area where there's lots of trash, a small searchcoil is best. These typically range in diameter from 9 to 14 cm (3.5" to 5.5"). An elliptical shape (versus round) provides slightly more coverage and slightly better target separation. Small searchcoils go nearly as deep as standard searchcoils (typically 20-25 cm diameter round or 25 – 30 cm length if elliptical)."

With this in mind I used an elliptical coil housing of 9.5" x 5" (Fig. 6-3), and wound a concentric coil with a 2" diameter inner loop. The intent was to combine the benefits of a small search loop with the coverage of the elliptical shape.

During nulling of the coil it was not possible to achieve an amplitude at the DISC preamp output of less than 1.16V, which equates to 58mV at the coil itself. It is suspected that this was due to the inner coil being relatively near the outer loop, despite only being 2" in diameter.

The frequency of the ringing was 10kHz with a parallel capacitance of 110nF. The outer loop had an inductance of 300µH, and the inner loop was 2.3mH. Like the Fisher coil, the PI main sample delay was 18µs. Strangely, the elliptical coil required the DISC sample pulse to be positioned at the positive-going zero-crossing instead of over the peak. On the bench the rejection of ferrous targets was excellent. In the test garden the first coin was detected without any problems, but the response to the second coin was somewhat *iffy*.

Fig. 6-4: Large Concentric Coil

Large Concentric Coil

Going to the other extreme, a 12" concentric coil was then constructed with a 6" diameter inner loop (Fig. 6-4). The outer loop had an inductance of 347µH, and the inner loop was 1.14mH. With a parallel capacitance of 220nF the ringing frequency was 10kHz. It was possible for the DISC preamp amplitude to be nulled as low as 10mV, and the PI sample delay was the lowest so far at 13μs. DISC sampling was at the first positive peak with a delay of 50μs.

In the test garden it was possible (perhaps surprisingly) to clearly detect both the first and second coins. In a practical sense though, there was no clear advantage over either the Troy or Fisher coils, with the clear disadvantage of being comparatively heavy.

Fig. 6-5: Small Concentric Coil

Small Concentric Coil

In an attempt to find the best compromise between the different coil characteristics, a small 7" concentric was constructed (Fig. 6-5) with an inner loop diameter of 3". The outer loop inductance was kept low at 306μH, and the inner loop inductance was 467μH. With a parallel capacitance of 470nF the ringing frequency was 10.7kHz. The DISC preamp amplitude was measured as 860mV, which was quite acceptable. The PI main sample delay was 15μs, which (again) was very good for coin shooting.

This small coil was able to pick out the first and second coins, and provide excellent ferrous rejection.

Fig. 6-6: Tesoro Coil

Tesoro Concentric Coil

Like the Troy concentric coil, Tesoro also sold 4-pin types with a low inductance TX loop. For this test I used a Tesoro 4-pin 9"x8" [slightly] elliptical coil (Fig. 6-6) with an outer loop of 982μH, and an inner loop of 15.36mH. As you can see, the outer loop inductance is lower than the Troy, and the inner loop inductance is much higher. With a parallel capacitance of 22nF, the ringing frequency was 8.7kHz.

The PI main sample delay for this coil was 24μs (3μs faster than the Troy), and the DISC preamp amplitude was -432mV at the first negative peak.

The performance of this coil in the test garden was comparable to the Troy.

It appears from testing the various coils above that Voodoo is capable of working with a wide range of loop inductances, and that the ringing frequency of the inner loop is completely unrelated to the TX pulse repetition rate of the PI transmitter, but is wholly dependent on the value of the parallel capacitance in association with the loop inductance. This fact was confirmed by attempting to tune the ringing frequency to 10kHz (to match the 1000pps TX rate), but there appeared to be no advantage whatsoever. This was also confirmed by the testing done with the Tesoro coil running at 8.7kHz. Many tests were made with the homemade coils (large, small and elliptical) to establish whether there was any advantage to setting the coil nulling to its minimum value, or to either side of the minimum. Offsetting the null phase of the coil is often a requirement for VLF detectors to maintain consistency with the ground balance and discrimination controls. For Voodoo there was really no advantage either way, and it was just simpler to set it to the minimum residual voltage. Of course it should be obvious that the final nulling point does determine the value of the DISC delay. Therefore this should be taken into account when balancing the coil.

As mentioned in Chapter 3, the DISC preamp circuit contains a trimmer that can adjust the gain between 19.6 and 39.2. Experiments showed that there was not a dramatic difference between minimum and maximum DISC gain for any of the coils. All of the measurements above were made with the lowest gain.

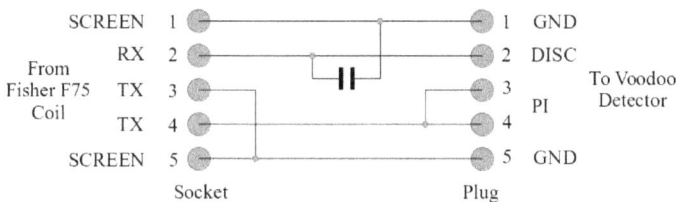

Fig. 6-7: Fisher F75 to Voodoo Adapter Cable

Coil	PI/TX/ Outer Loop	DISC/RX/ Inner Loop	PI Main Sample Delay	Parallel Capacitance	Frequency	DISC Delay
Troy	1.12mH	6.96mH	26μs	33nF	10.5kHz	45μs
Fisher F75	696μH	6.9mH	18μs	33nF	10.5kHz	45μs
Elliptical	300μH	2.3mH	18μs	110nF	10kHz	50μs
Large	347μH	1.14mH	13μs	220nF	10kHz	50μs
Small	306μH	467μH	15μs	470nF	10.7kHz	45μs
Tesoro	982μH	15.36mH	24μs	22nF	8.7kHz	60μs

Table. 6-1 : Coil Parameters and Detector Settings

Chapter 7 _____ *Conclusion*

"A conclusion is the place where you got tired of thinking."

--- Arthur McBride Bloch – American writer, author of *Murphy's Law* books

The purpose of this book was to document the development and construction of the Voodoo hybrid (pulse induction / continuous wave) metal detector. Voodoo is an experimental device, and for that reason has not been presented here as a fully fledged and polished project. That was never the intention. The menu system, for example, allows the operator to adjust parameters that the majority of users would find totally confusing. Also, the PI and DISC thresholds would really need to be linked together and made adjustable from an external potentiometer, and not buried in the menus. Apart from that the detector is very stable in operation once the correct settings are found for a particular coil, and it is able to pick out non-ferrous targets buried amongst ferrous trash.

Unlike a VLF detector with a visual display indicator (VDI), Voodoo's display acts like an analog meter. When a target is under the coil, the display moves either right (non-ferrous) or left (ferrous) depending on the target type. If the target is a mixture of both non-ferrous and ferrous, the meter will indicate whether the response is predominately of either type, and provides a much clearer indication than a VDI. In this scenario most VDIs will either bounce around between the various target types, whereas other more sophisticated detectors may provide a graphical display in the form of a histogram. With Voodoo the number of bars on the display (either side of the centre line) shows the proportion of non-ferrous to ferrous. For many detectorists a VDI is nothing more than a gimmick, as it is well known that target identification is very much a guessing game. The accuracy of the VDI numbers can vary wildly once the target is buried in the ground matrix, and are also dependent on orientation. Voodoo only attempts to discriminate between non-ferrous and ferrous, which improves the ability to separate the desirable targets hiding amongst the undesirable iron trash. Figures 7-1 and 7-2 provide examples of the indication from a non-ferrous and a ferrous target. In both figures the LCD is displaying 5 bars, but these can range from 1 to 7. Also, note how the battery symbol changes depending on the battery voltage. When Voodoo is working in PI mode the LCD displays "PULSE" instead of "HYBRID", and (of course) does not provide ferrous/non-ferrous discrimination, but responds to all metal targets regardless of type. Also, the meter function is inactive, and only displays the separator bar on the 2nd line.

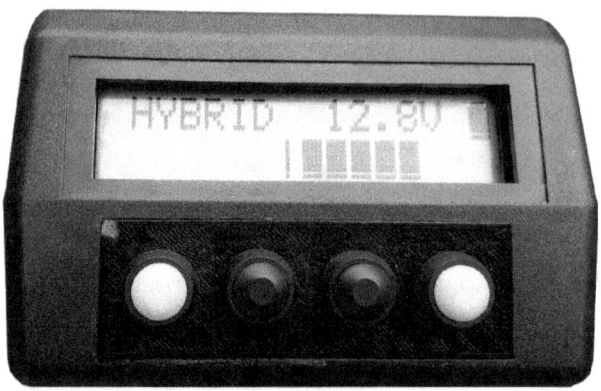

Fig. 7-1: LCD Indication of a Non-Ferrous Target

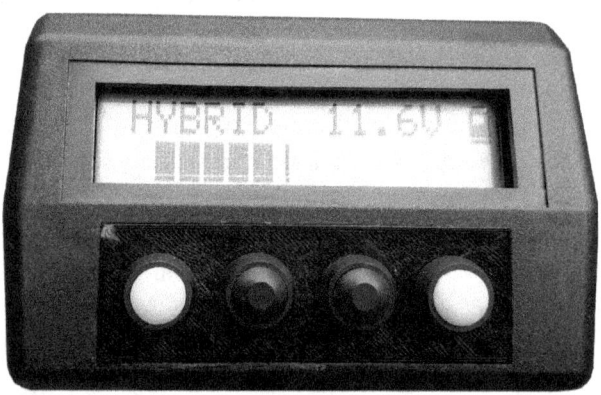

Fig. 7-2: LCD Indication of a Ferrous Target

The general concept of using a pulsed transmitter to stimulate a receive coil, causing it ring at a specific frequency, is not unknown, and a similar method has been used in some Heathkit and Radio Shack designs. In Chapter 8 of ITMD a simple transmit-receive (TR) detector was presented where the transmitter produced a series of pulses that resulted in a set of decaying sine waves. The Voodoo approach, however, is somewhat different because the transmitter is critically damped, and only the receive coil is set into resonance at a frequency that is solely defined by the inductance of the receiving loop and a parallel capacitor. This method has in fact been *rediscovered* on numerous occasions by different people, but no-one (to my knowledge) have ever been successful in combining this technique with a PI detector to provide ferrous/non-ferrous discrimination. It appears that other attempts have failed because the detector was severely affected by the ground matrix, in the same way as all TR detectors. This was also the case with the ITMD TR detector, and in fact it was the introduction of motion detectors that heralded the death knell for the TR. Most experimenters who *rediscovered* this technique have had a background in pulse induction, and therefore were used to using non-motion with a self-adjusting threshold (SAT). Consequently, the problem of ground effect was extremely troublesome and experiments were often abandoned for that reason.

Voodoo is subtly different in that it has two separate preamps, one for the PI channel, and a second preamp for the receive coil. Although the PI channel is unaffected by non-mineralized (or even moderately mineralized) soil, the second (DISC) channel has the same issues as the old TR designs. A non-motion configuration was tested with Voodoo and found to be practically unusable. The use of a SAT circuit has the same effect as using a single-differentiating architecture, where the detection of the target starts at the edge of the coil and terminates at the centre. On the bench this topology works well with good discrimination, but in the real world it acts like a widescan coil, and this results in a slow recovery and makes it difficult to separate targets that are close together.

A double-differentiating architecture was the final solution that allowed this idea to work successfully. The only drawback is that the depth may be less than people are used to with a standard PI, as the detector now responds to the (rate-of-change)^2 of the target signal, rather than the signal itself. However, without double-differentiation (and hence ferrous/non-ferrous discrimination) detecting in areas with large amounts of ferrous trash would be particularly onerous. A similar situation exists with PI detectors that provide ground balance. The extra GB sample subtracts a portion of the target signal and hence some depth is lost, but without ground balance it would be impossible to detect in areas where mineralization is very high.

To demonstrate the difference between non-motion, single-differentiation, and double differentiation, an LTspice simulation was created. The schematic is shown in Fig. 7-3, and the plot results are in Fig. 7-4.

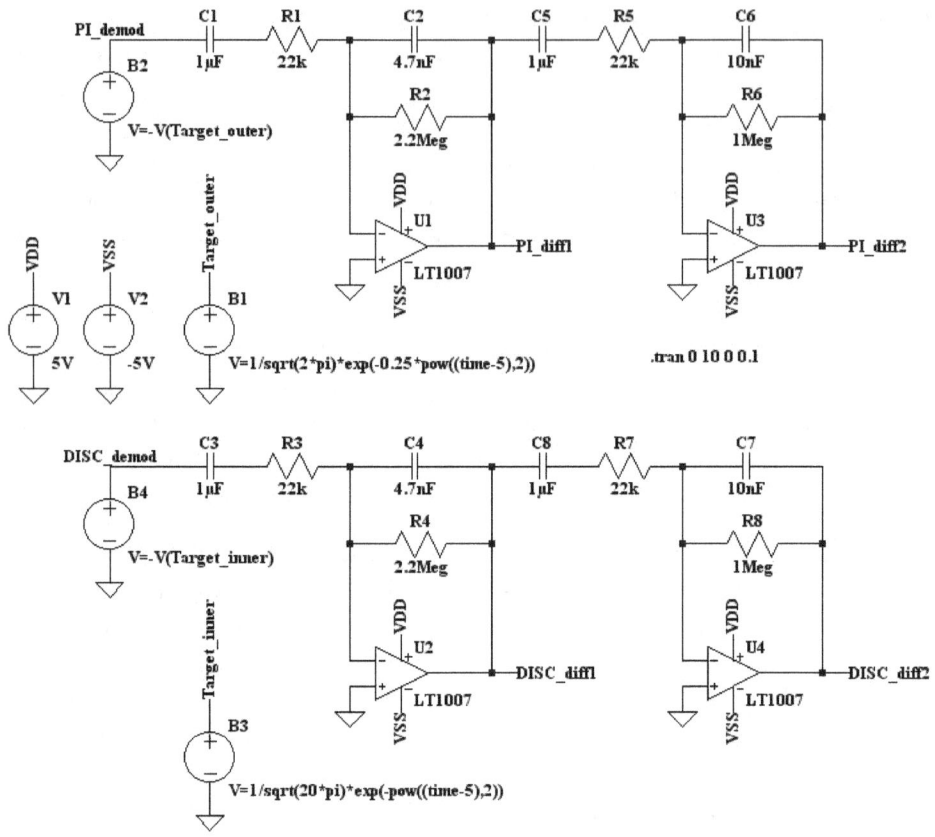

Fig. 7-3: LTspice Simulation of Target Response

The top part of the schematic (Fig. 7-3) represents the PI channel, and the bottom part represents the DISC channel.

Behavioural voltage source B1 generates a normal (bell curve) distribution to represent a metal target passing over the coil, and the top pane in Fig. 7-4 shows the target signal received at the outer (PI) loop. As you can see, the target signal starts at the edge of the outer loop, increases as it approaches the centre of the coil, and then decreases again towards the other edge. Since the preamp in Voodoo inverts the signal from the coil, the output from the synchronous demodulator will also be inverted from that shown in the

The Voodoo Project

top pane. The behavioural voltage source B2 provides this inversion before being fed into the PI filter circuits. The middle pane of Fig. 7-4 shows the result after passing through the first differentiating filter. In this case the signal initially rises as it responds to the target passing over the coil. As the signal approaches its maximum rate-of-change the output signal from the filter circuit slows down, until at maximum target signal the rate-of-change becomes zero. As the target signal starts to decrease, the filter output then goes negative until after passing the point of maximum negative rate-of-change it returns to zero.

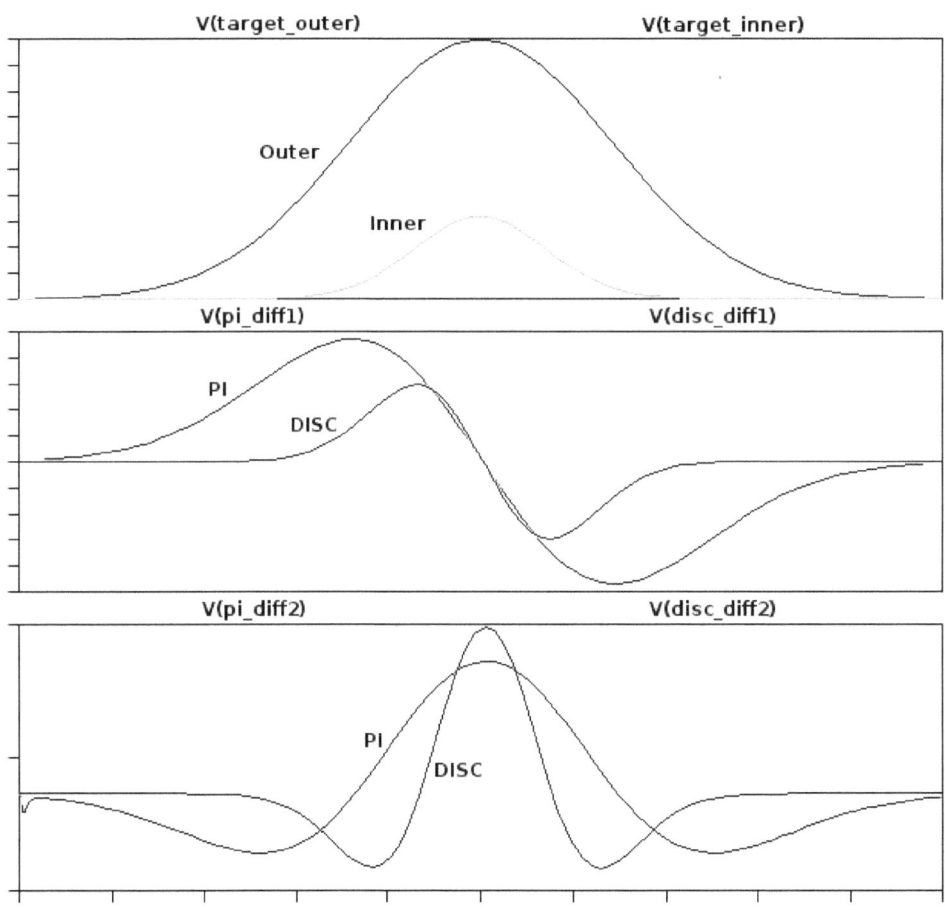

Fig. 7-4: LTspice Target Response Results

If the detector were to use the output of the first filter for target detection, the target would trigger a response as it approached the edge of the coil, and stop when it reached the centre.

Note that the DISC channel reacts the same as the PI channel, with B4 generating the target response at the inner loop and B3 providing the signal inversion of the preamp. The important difference between the two is that the PI channel gives a positive response for all targets, whereas the DISC channel goes positive for a non-ferrous target and negative for ferrous. The plots in Fig. 7-4 show the target response for a non-ferrous target. Therefore, when both the PI and DISC channels are above the threshold values, the detector gives an audible response. However, when a ferrous target is detected, the DISC channel is below the threshold and there is no audible signal generated.

Notice (from the middle pane of Fig. 7-4) that there is an additional problem associated with attempting to compare the PI and DISC channels' single-differentiated outputs to provide target discrimination. You can readily see that the peaks of the PI and DISC waveforms are not aligned, so that the PI channel has already passed its point of maximum signal when the DISC channel starts to respond. This problem was resolved by adding a second differentiating filter, providing so-called double-differentiation.

An examination of the bottom pane in Fig. 7-4 shows clearly that the point of maximum signal for both channels are now temporally aligned. This point occurs at the centre of the coil. In practice the target appears to give an audible response as it leaves the centre of the coil, rather than as it approaches the edge. This gives the advantage of sharpening the response of the detector to small targets such as hammered silver coins for example. When you compare the shape of the curves in the bottom pane with those in the top pane, it becomes obvious why target separation is also much improved.

One important thing to note from Fig. 7-3 is that the final filter stage inverts the signal from the first stage. At first this may seem to be a bit odd, but examining the waveforms in the bottom pane of Fig. 7-4 will make things clearer. The response, after double-differentiating has taken place, now has 3 peaks. Without signal inversion there would be 2 beeps for every non-ferrous target, and no signal as the target passed over the centre of the coil.

The PI channel inherently ignores any response from neutral (or even moderately mineralized) soil, since any eddy currents will have completely died away by the time the preamp output is sampled. In the case of the DISC channel, the double-differentiating filters allow it to only respond to fast-changing signals and ignore any slow-changing signals due to the ground matrix. However, it must be made clear (as mentioned previously) that Voodoo was never designed to operate in areas of high mineralization. The primary goal of the project was to create a hybrid detector that could be used inland, where a standard PI could not operate due to the sheer amount of ferrous trash in the ground.

Areas that require further research and development are:

1. Determination of the optimum coil parameters, such as loop inductance and physical construction.

2. Adding a ground balance facility to extend the detector's capability to work in areas of higher mineralization.

3. Optimization of the filters (most possibly implemented in software) to further improve ferrous rejection.

4. Simplifying the menu system to remove any parameters that were only there for ease of adjustment during development and testing.

5. Anything else I haven't thought of.

Hopefully all of this has made some sense … since (as the quote at the start of this chapter states) ... *"I've got tired of thinking"*.

Appendix A

Component Parts List

"To estimate the time it takes to do a task, estimate the time you think it should take, multiply by 2, and change the unit of measure to the next highest unit. Thus we allocate 2 days for a one-hour task."

--- Westheimer's Rule

ICs

Component	Names	Quantity
16x2 LCD Module	U4	1
78L05 +5V regulator	U7	1
79L05 -5V regulator	U6	1
4066BP bilateral switches	U5	1
LT1054 voltage converter	U8	1
NE555 timer	U11	1
NE5532 dual opamp	U3	1
PIC18F4520 microcontroller	U2	1
TL071 single opamp	U1	1
TL072 dual opamp	U9-10	2

Diodes

Component	Names	Quantity
1N4148	D2-7	6
MUR460	D1	1

Capacitors

Component	Names	Quantity
1μF	C15-18, C33	5
4n7	C19-20, C24	3
10nF	C25, C30	2
22pF	C2, C4	2
22μF electrolytic	C8, C28	2
47μF electrolytic	C3, C6, C9	3
100nF	C5, C7, C22-23, C26-27, C31, C34-35	9
100μF electrolytic	C13-14, C29, C32	4
470nF	C10-12	3
1000μF electrolytic	C1, C21	2

Transistors

Component	Names	Quantity
2N3904 NPN	Q2, Q4, Q6-8, Q10, Q12	7
2N3906 PNP	Q3, Q5, Q9	3
FQT7N10 mosfet	Q11, Q13	2
IRF840 mosfet	Q1	1

Connectors

Component	Names	Quantity
2-pin 100th pitch	PL1-6, PL10-13	10
3-pin 100th pitch	PL7-8	2
5-pin ICSP	PL9	1

Resistors (1%)

Component	Names	Quantity
1k	R8, R19, R25, R37 R57, R62	6
1M	R18, R58, R63-64	4
1R0	R6, R53	2
2M2	R52, R54-55	3
3R3 2W	R2	1
5k1	R15	1
10k multi-turn trimmer	R41	1
10k	R1, R3-5, R7, R9-13, R23, R26, R28, R33-36, R39-40, R42-43, R45, R49-50, R56, R60-61, R67	28
12k	R68	1
20k	R24, R48	2
22k	R27, R31, R44, R46-47, R51, R59, R65	8
27R	R17, R32	2
33k	R14, R20	2
33R	R66	1
100k multi-turn trimmer	R22, R30	2
100k	R16, R21, R29	3
100R	R38	1

Other

Component	Names	Quantity
2-pin jumper	JP1-3	3
Test point	TP1-20	20
20MHz crystal	XTAL1	1

Miscellaneous

Depending on the actual construction, there are some additional components required beyond the LCD display and those on the PCB itself.

For example:

- On-off switch (SPST)

- Speaker and/or headphones (8-64 ohm)

- Headphone socket (recommended)

- IC sockets (optional)

- Electronic control box

- Push-buttons for menu system

- Battery pack (10x NiMH AA)

- Battery charger socket

- External ICSP connector

- Coil connector (5-pin)

- Coil (either commercial or custom made)

- Coil cable (preferably coax, RG58 or equivalent)

- Detector stem, arm cup, stem bolts, etc.

- Damping resistor

- Resonant capacitor for RX loop

- Battery pack charger (see Appendix B)

Appendix B

Battery Pack Charger

"Hell! there ain't no rules around here! We are tryin' to accomplish somep'n!"

--- Thomas Edison (1847 – 1931)

As explained previously, the battery pack consists of 10x AA NiMH non-removable rechargeable batteries, and hence these need to be charged in situ. This appendix describes a suitable charger circuit that (for the 2500mAh capacity used) is able to charge the pack from fully discharged (1V per cell, i.e. 10V) within about 12 hours. The easiest and cheapest way to charge NiMH batteries is to charge them at a rate of C/10, which is equivalent to 10% of the rated capacity per hour. In this particular case the charge rate needs should be no greater than 250mA, which removes the need for an end-of-charge sensor.

Circuit Description

The charger is designed around an LM317, which is 3-terminal adjustable regulator, and is configured as a precision current source.

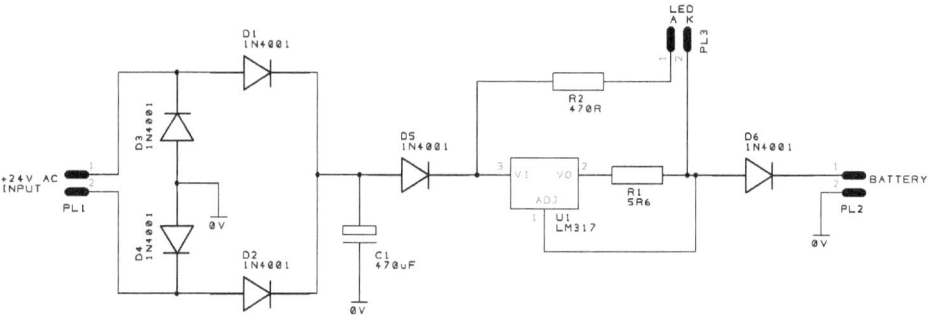

Fig. B-1: Charger Schematic

With reference to Fig. B-1, the regulator will provide a constant current output according to the following equation:

$$I_{out} = \frac{V_{ref}}{R_1} = \frac{1.25}{5.6} = 223mA$$

There is an LED indicator that only illuminates when the battery is charging, and this also contributes some current to the battery.

For example (in practice) the following measurements were made:

Battery Voltage	I_{R1}	I_{LED}	I_{CHG}
10V	222mA	19mA	241mA
13.5V	222mA	12mA	234mA
0V (short-circuit)	221mA	39mA	261mA

As you can see from the above table, the charging current is a nominal 240mA. In this case, the 2500mA battery pack would be charged in 2500/240 = 10.4hrs, assuming no charging losses. In reality, it takes around 12 to 15 hours to reach full charge.

The input voltage to the regulator must be greater than:

$$V_i = V_{bat} + V_{ref} + V_{headroom} = 13.5 + 1.25 + 3 = 17.75V$$

There is an additional diode at the input to the regulator, so that the voltage required at the output of the bridge diode arrangement must be greater than 18.45V. Clearly there is sufficient headroom provided by the mains transformer, which has an output of 24V. The transformer was not included on the PCB, as there are numerous different formats for these devices. There are many transformers available with two 12V outputs that can readily be connected in series to achieve the required 24V.

Note that there is a second diode connected between the circuit and the battery pack. This diode prevents any current feeding back from the battery to the charger.

A suggested layout for the PCB is shown in Fig. B-2 as a 3D view. The component placement is provided in Fig. B-3, and Fig. B-4 shows the copper tracks (as viewed through the board from the top layer. This is a single-layer PCB with components on the top layer, and copper on the bottom.

A mounting hole is provided so that the regulator may be fitted with a heatsink.

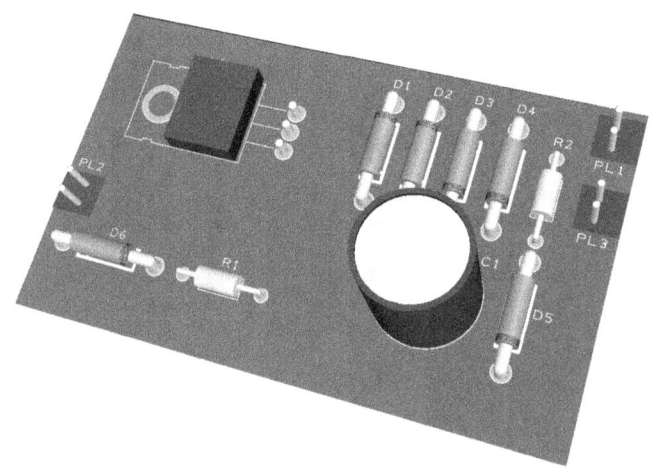

Fig. B-2: 3D view of Charger PCB

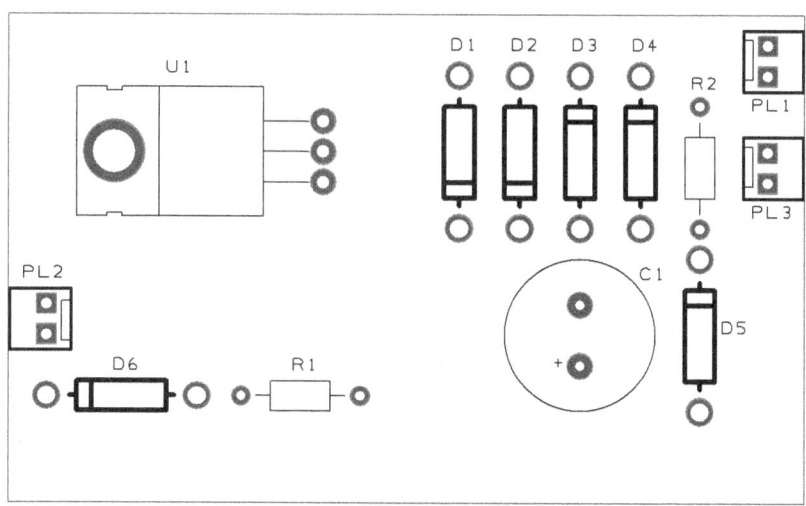

Fig. B-3: Component Placement

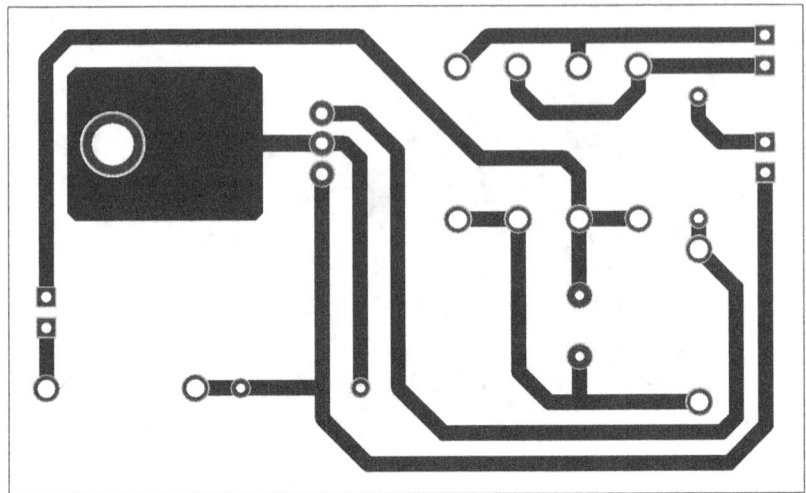

Fig. B-4: Copper Tracks - Bottom Layer (looking through the PCB)

Appendix C

An Investigation into "Flat-topping"

"Engineer … noun [en-juh-neer] Someone who does precision guesswork based on unreliable data provided by those of questionable knowledge."

--- T-shirt slogan

There was a paper published by F. B. Johnson in 1956 [1] (end of page 5 / start of page 6) where he discusses the suggestion by J. H. Westcott to use the energy from a bank of capacitors to provide the current for the magnetic field in a pulsed bomb locator. In this section he also points out that the TX-on time needs to be approximately ten times longer than the longest eddy current decay time encountered:

"If the eddy currents have not decayed to approximately zero at the time of switch-off, there will be a loss of signal, because the switch-off induces currents in the opposite direction to those induced by a switch-on. In the case of a rapid switch on and off the eddy currents are exactly cancelled out, and hence there is no signal to observe."

This is theoretically correct, but in reality the "*ten times longer*" conclusion is somewhat pessimistic, and 5τ (following the Law of Diminishing Returns) sufficiently long enough to eliminate the problem. However, what this indicates is that the issue of allowing eddy currents induced in a metal target at switch-on to decay fully before turning off the coil current was already known during the 1950s.

It has been suggested by others that even a time of 3τ can be used in practice, and will be close enough without wasting power to get the last few percent of target signal. This can clearly be seen from Table C-1.

The table results were calculated to two significant places from:

$$I_L = I(1 - e^{\frac{-t}{\tau}})$$

The term "flat-topping" refers to the practice of allowing the coil current to reach 100%, effectively flattening out, with the result that the rate-of-change of current equals zero.

During this time, the eddy currents in the target need to have reached a maximum, and the coil current needs to remain "flat-topped" long enough for these eddy currents to decay to zero. If there are still eddy currents flowing in the target at the point where TX-off occurs, the reverse eddy currents generated by the collapsing magnetic field will be reduced, thus weakening the target response at the receiver.

Number of target tau (τ)	Eddy current decay (%)
1	63.21
2	86.47
3	95.02
4	98.17
5	99.33
6	99.75
7	99.91
8	99.97
9	99.99
10	100.00

Table. C-1: Number of target τ versus eddy current decay

The question is: "How important is this really in practice?".

To answer this question we can run a SPICE simulation on a typical PI front-end, and examine what happens to the eddy currents in the target, and the subsequent response at the output of the preamp.

With reference to Fig. C-1:
The simulation is configured to run a parametric sweep of the value of the series resistor (res) and the TX pulse width (pw). Using the *table* directive in LTspice, the simulation first runs with the series resistor set to 3R3, and the pulse width at 50 μs. Since the coil inductance is 300μH, and the total dc resistance during the mosfet on-time is 6R0 ($R_L = $ 2R7, and $R_S = $ 3R3) then the tau of the coil is:

$$\tau = \frac{L}{R} = \frac{300\mu}{6} = 50\mu s$$

Therefore the TX pulse width in the first simulation is equal to the tau of the coil, and (from Table. C-1) it can be seen that the coil current will rise to 63.21% of the maximum value.

The parallel combination of L_2 (100μH) and R_3 (10R) are used to represent a target with a decay constant of:

$$\tau = \frac{L}{R} = \frac{100\mu}{10} = 10\mu s$$

The second simulation of the parametric sweep changes the series resistor to 6R86, and the TX pulse width to 250μs (which is equal to 5τ, or 99.33% of the maximum possible current). Note that the values for the series resistor have been chosen so that the current flowing in the coil at the point of switch-off will be substantially the same in both cases.

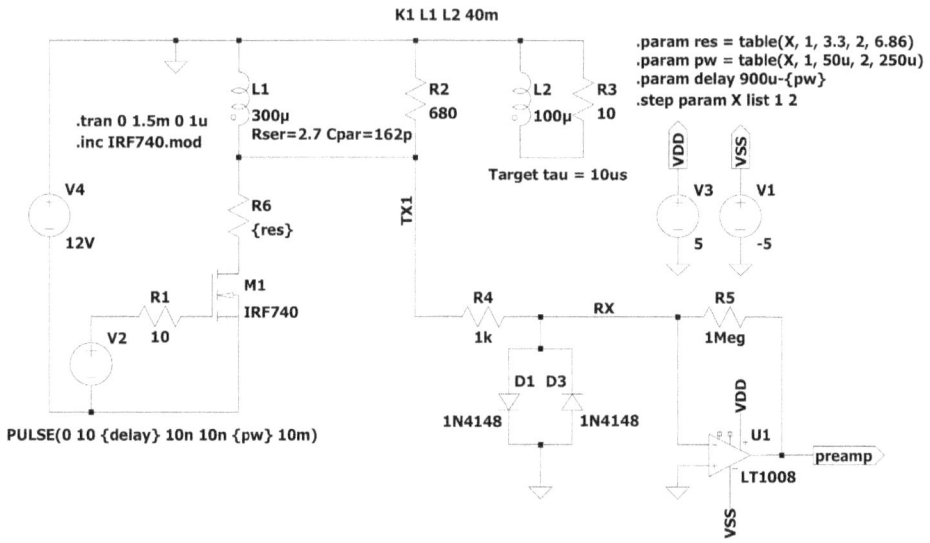

Fig. C-1: Flat-top simulation

The eddy current response of the target is given by:

$$I = e^{\frac{-Rt}{L}}$$

Fig. C-2 shows the plot results of the parametric sweep. From pane 3 it can be seen that the eddy currents in the target do not have sufficient time to decay to zero before the mosfet is switched off in the case where the TX pulse width is 50µs. On the other hand when viewed from the perspective of the preamp output, the sensitivity to the target is similar for both cases and hardly seems worth the effort.

If we now run another parametric sweep , but only vary the TX pulse width, and keeping the series resistance constant at 3R3, the plot results can be seen in Fig. C-3. In this particular simulation the difference is much clearer. The main reason, of course, is that the simulation run with the 250µs TX pulse width does not have its maximum current artificially restricted to the same value as that for the 50µs pulse width.

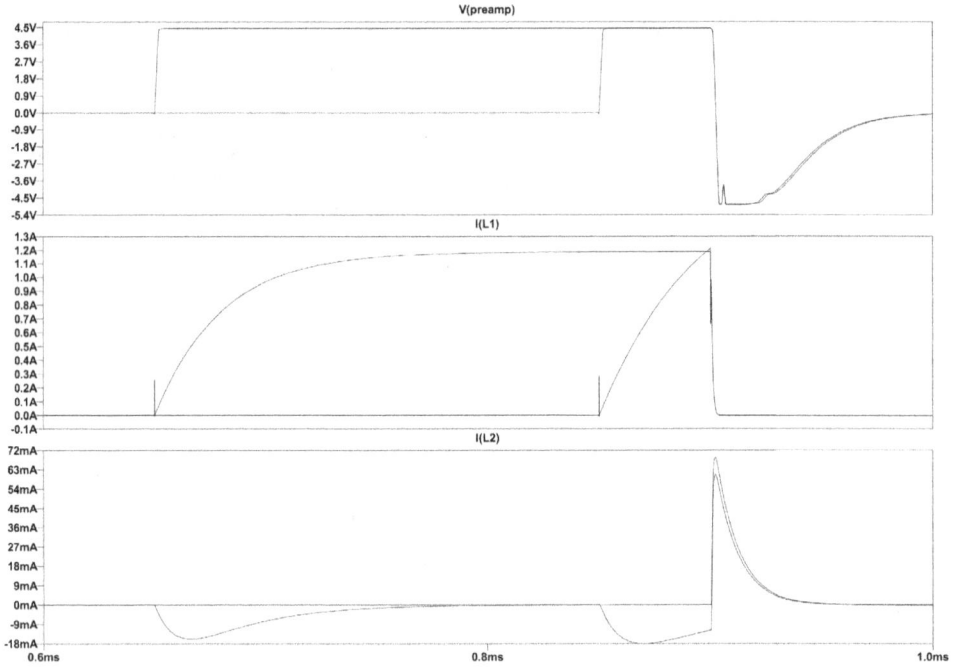

Fig. C-2: Plot results from parametric sweep of R$_s$ and TX pulse width

But this is not the end of the story. The results that can be achieved by flat-topping the coil current, are also affected by the target time constant. In both Figures C-2 and C-3, the target was fully stimulated during switch-on, and the eddy currents have had some time to decay, even though they did not decay to zero. If we increase the target tau to 50μs and use the same parametric sweep parameters as in the first example (see Fig. C-4) we get the plot results that are shown in Fig. C-5.

As you can clearly see, the 50μs target only has sufficient time for the eddy currents to increase to maximum, and there is no time for these to decay when the TX pulse width is set to 50μs. It can be concluded from the simulation results that "flat-topping" is not a major factor in all situations. In fact, there are many instances where simply ignoring "flat-topping" can provide better results. Much depends on the type of target you are planning to detect, the coil parameters, main sample delay, TX pulse width, and the capacity of the battery pack.

For a standard PI design with a 300μH coil having 2R7 dc resistance, and a 3R3 series resistor, the coil tau is 50μs. With a typical TX pulse width of 100μs, this represents only

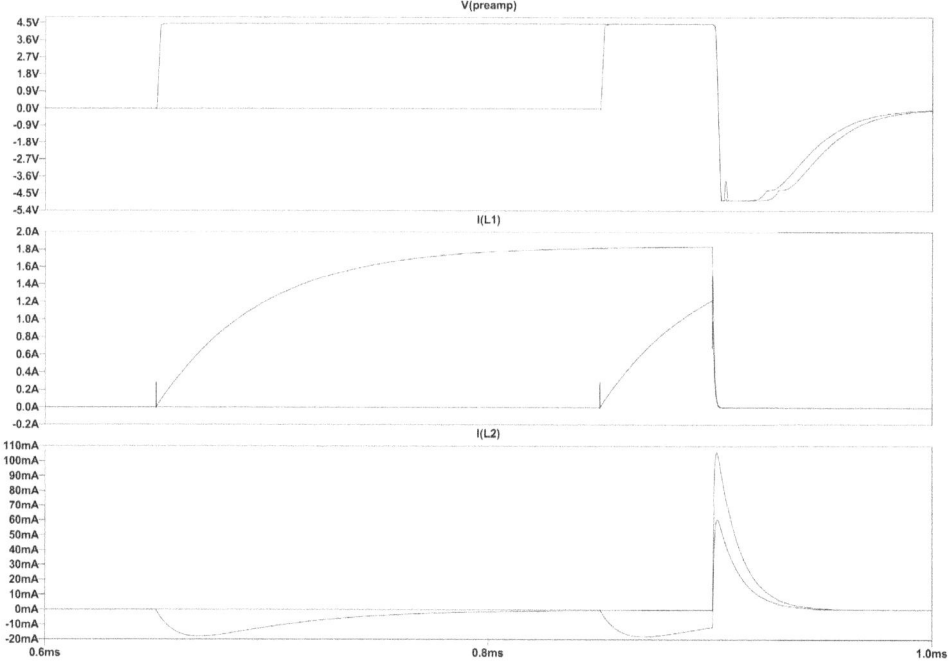

Fig. C-3: Plot results from parametric sweep of TX pulse width with R$_s$ constant

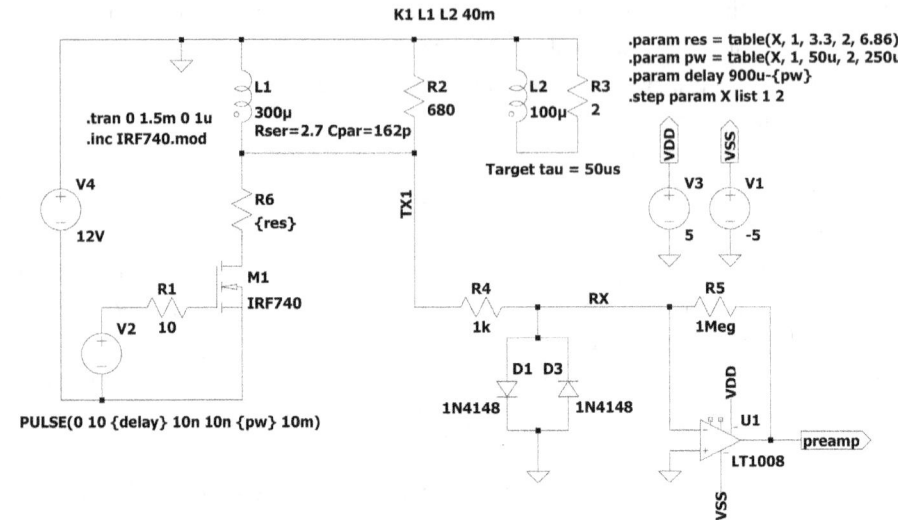

Fig. C-4: Flat-top simulation with 50µs target decay constant

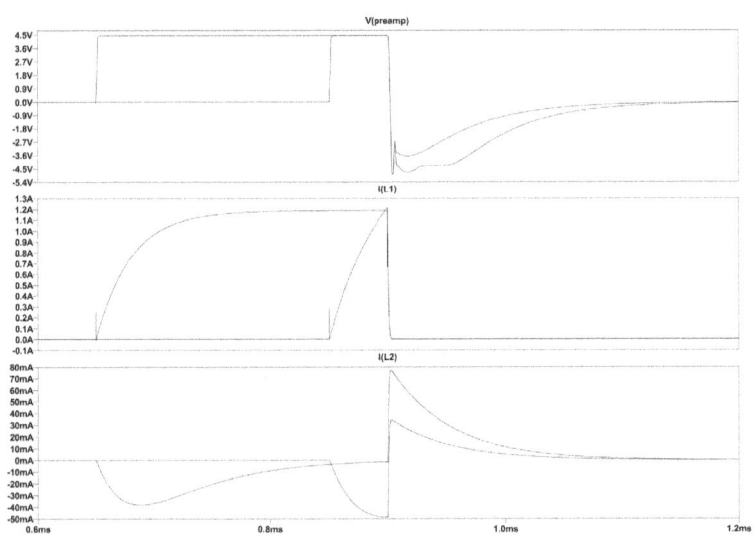

Fig. C-5: Plot results from parametric sweep with target decay constant set to 50µs

2x the coil tau. In this example, any targets with a tau above 10µs will contain residual eddy currents at switch-off. Despite this, such a standard configuration performs very well in the field, although increasing the TX pulse width (and lowering the TX pulse rate) can provide improved depth on targets with higher decay constants. Not only because of the higher maximum coil current, but also because of flat-topping.

As always, engineering is a matter of compromise.

 **Preamp output oscillations**

You may have noticed in the simulation results that there are some small oscillations in the preamp output signal. For the purposes of the simulation, I calculated the damping resistor using the formula:

$$R_d = \frac{1}{2}\sqrt{\frac{L}{C}} = \frac{1}{2}\sqrt{\frac{300\mu}{162p}} = 680$$

Although approximately correct, it does not take into account the mosfet capacitance (Coss), and hence the circuit is not perfectly damped. The effects of the mosfet capacitance can be reduced by inserting a diode in the path to the coil. This subject is explored in more detail in Appendix D.

If you run the simulations in Fig. C-1 and Fig. C-4, you will notice the assumption that setting the TX-on time to 5 times the target tau to allow the eddy currents to decay by 99.33% of the maximum value is not exactly correct. When the mosfet is turned on, the target eddy currents are non-existent, and so must build up exponentially from zero before starting to decay. Although the initial build-up of eddy currents is proportional to the tau of the coil and the decay is dependent on the tau of the target, analysis is further complicated by the fact that the time taken for the eddy currents to reach their peak is dependent on many other parameters as well, and not simply related to the coil tau. There are innumerable factors to take into account when designing a PI detector, and flat-topping may not be so important as you may initially think, depending on the application.

References: 1. F.B. Johnson M.A. - A Pulsed Bomb Locator (1956)

Appendix D

Series Diode in PI TX Circuit

"As far as the laws of mathematics refer to reality, they are not certain, and as far as they are certain, they do not refer to reality."

--- Albert Einstein (1879 - 1955), "Geometry and Experience", January 27, 1921

The majority of PI designs do not include a diode in series with the coil, so why put one there, and what is its purpose?

Initially when I first heard of the idea of putting a diode (often an MUR460) in between an IRF740 mosfet and the coil to allow for earlier sampling, there were two possibilities that came to mind:

1. The diode capacitance is in series with the output capacitance of the mosfet, and hence the overall capacitance is reduced, allowing earlier sampling.
2. The breakdown voltage of the MUR460 is 600V, which is above the IRF740's V_{DS} of 400V. With these two devices in series a higher flyback voltage would be possible before breakdown occurred, resulting in a faster settling time.

Increasing the TX-on time such that the flyback voltage was above 400V showed clearly that option 2 was incorrect, as breakdown still occurred at 400V. The diode is forward-biased during the TX-on period and the coil current increases exponentially. Once the mosfet is turned off the coil current starts to decrease, and the coil voltage reverses polarity and increases dramatically in amplitude. At this time the diode becomes reverse-biased.

So what about option 1?

From simulation it is clear that the mosfet output capacitance appears to get partially disconnected from the coil by the diode. You can see that by examining the shape of the coil current after TX-off. There is a small hump in the decay curve where the coil's parasitic capacitance initially becomes charged and then subsequently discharges. This hump is reduced in amplitude when the diode is present. Also, the amount of reduction is related to the ratio between the mosfet's output capacitance and the coil's parasitic capacitance.

If you attempt to simulate the difference produced by a series diode, then you need to be aware that some mosfet subcircuits provide modelling for the nonlinear variation of C_{oss} as a function of drain to source voltage V_{DS}, whereas some others do not. The value quoted in the mosfet datasheet is specified at 25V which is not that useful for our purpose, as it is more useful to know the value of C_{oss} effective which is defined as a fixed capacitance that would give the same charging time as the output capacitance of a mosfet while V_{DS} is rising from zero to 80% V_{DS} with $V_{gs} = 0V$.

The simulation in Fig. D-1 shows three test circuits (one for each of the mosfets), and the results are shown in Fig. D-2. If you measure at the 80% point (400V for left, 640V for middle, and 320V for the right-hand circuit) t_c is 27.9μs, 23.1μs, and 33.9μs respectively. Then using:

$$C_{oss} = \frac{-t_c}{R * ln(1 - \frac{V_c}{V_{DS}})}$$

where: t_c = time to reach 80% V_{DS} with $V_{gs} = 0$
V_c = voltage at time t_c
V_{DS} = breakdown voltage of mosfet
R = series drain resistor

we have:

$$C_{oss} = \frac{-27.9\mu}{100k * ln(1 - \frac{400}{500})} = 173pF \text{ (left circuit)}$$

$$C_{oss} = \frac{-23.1\mu}{100k * ln(1 - \frac{640}{800})} = 144pF \text{ (middle circuit)}$$

$$C_{oss} = \frac{-33.9\mu}{100k * ln(1 - \frac{320}{400})} = 211pF \text{ (right circuit)}$$

Note that the voltage sources are set to the maximum voltage for each mosfet. If you raise these voltages to 600V, 900V, and 500V respectively, you will readily see that the avalanche voltage is also modelled correctly.

The Voodoo Project

After running a number of simulations with and without the diode, the [theoretical] conclusion is that the diode would be useful / noticeable when the value of C_{oss} is larger than the coil's capacitance, which is often not the case for *quick and dirty* home-made coils above 300µH.

Of course there's little point playing around with the diode anyway if your mosfet is going into avalanche mode. Then it's more important to either replace it with a higher V_{DS} mosfet, or reduce the TX-on time to prevent it from avalanching.

From an initial inspection this implies that the diode only eliminates about 50% of the mosfet capacitance. Also, for each simulation you need to set R_d for critical damping. If you leave R_d unchanged, the simulation results will indicate that adding the diode actually makes things worse.

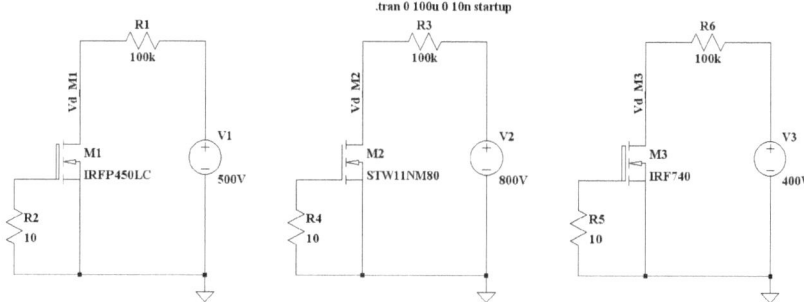

Fig. D-1: Mosfet C_{oss} measurement schematic

The final conclusion (at least from a simulation perspective) is that the series diode does improve the ability to early sample, but how much you gain depends on a number of factors.

More food for thought …

Here's some tests and measurements made on a real PI board using an IRF740 mosfet without a series diode, and comparisons made with simulation:

I used a mono coil with an inductance of 432µH and a dc resistance of 2R6. The damping resistor tool (see ITMD, end of Chapter 11) was used in conjunction with an oscilloscope connected to the preamp output to find the value at critical damping. This was found to be

540R with the diode shorted out. However, the actual point where critical damping is achieved is somewhat subjective, so I do not consider this method to be very accurate where its purpose is to determine the total capacitance in the TX circuit.

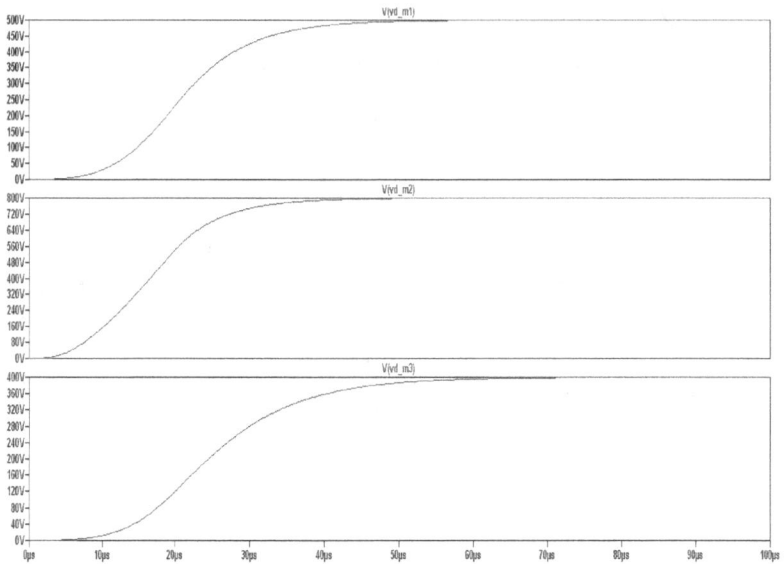

Fig. D-2: Mosfet C$_{oss}$ plot results

Using the damping resistor value of 540R we can determine the total capacitance as:

$$C_T = \frac{L}{4 * R_d{}^2} = \frac{432\mu}{4 * 540^2} = 370pF$$

where this represents the coil, cable, mosfet, and other parasitic capacitances.

A better method would be to remove the damping resistor altogether, and measure the resonant frequency of the resultant ringing. In this case without the series diode in circuit the resonant frequency was measured as 438.6kHz, which allows us to easily calculate the total capacitance:

$$C_T = \frac{1}{\omega^2 * L} = \frac{1}{(2\pi * 438.6k)^2 * 432u} = 305pF$$

As you can see there is a discrepancy between the two methods used. Since method one is somewhat subjective, we will continue with method two which is based on measuring the resonant frequency.

In this case the damping resistor value can be calculated from:

$$R_d = \frac{1}{2}\sqrt{\frac{L}{C}} = \frac{1}{2}\sqrt{\frac{432u}{305p}} = 595R$$

With the series diode connected the resonant frequency was measured as 602.4kHz, and following the same procedure as before:

$$C_T = \frac{1}{\omega^2 * L} = \frac{1}{(2\pi * 602.4k)^2 * 432u} = 162pF$$

and:

$$R_d = \frac{1}{2}\sqrt{\frac{L}{C}} = \frac{1}{2}\sqrt{\frac{432u}{162p}} = 816R$$

From the above it is clear that the diode provides a reduction in total capacitance allowing the use of a higher value damping resistor. Also, the reduction can easily be calculated as 143pF. Since we determined earlier that C_{oss} effective of the IRF740 mosfet is 211pF, this indicates that the diode effectively removes 67.8% of the mosfet capacitance from the TX circuit, bearing in mind that the capacitance due to everything else is unaffected by the presence of the diode.

With no diode, the total capacitance is 305pF. Therefore the capacitance associated with the coil, cable, and other parasitics is 305pF - 211pF = 94pF.

Inserting these values into an LTspice simulation shows a very small improvement in the lowest possible sample delay with the diode connected. This measurement was confirmed with a real PI circuit, although it was extremely difficult to see any significant improvement.

My conclusion is that inserting the diode doesn't make things worse, but (depending on many factors) there could quite likely be minimal improvement. Practical measurements do however show that the diode *blocks* the mosfet capacitance, resulting in a higher resonant frequency, and the requirement to increase the value of the damping resistor. In cases where you're trying to sample as early as possible, and "every little helps", then a series diode can only improve things.

The presence of the diode's depletion capacitance when reverse-biased is most likely the reason why it is able to block the mosfet's C_{oss} since it is effectively in series with C_{oss}. This would also explain why it is not able to provide 100% blockage. The diode used in both the practical and simulated circuits was an MUR460.

It is then simple to calculate that the depletion capacitance of the series diode while in reverse-bias must equal 100pF. With diode capacitance = 100pF, and C_{oss} effective = 211pF, the total series capacitance becomes 68pF.

Also, mosfet C_{oss} - series capacitance (with diode) = 211pF - 68pF = 143pF.

So the question now is whether a value for the depletion capacitance of 100pF is reasonable for an MUR460 under these operating conditions?

In the On Semiconductor datasheet for the MUR460, the diffusion capacitance is shown in the graph as 36pF at 50V, and over 100pF at the lowest voltage. The problem here is that both the mosfet C_{oss} and the diode diffusion capacitance vary non-linearly with applied voltage. Unless we want to solve the full equations, we can only estimate some of these values. At the moment, SPICE simulation agrees more or less with my real world measurements, and my conclusion remains as before.

Anyway, in a rather long-winded fashion, this is the reason why Voodoo contains a diode in series with the coil … just in case you were wondering.

Determining Target Time Constants

"Basic research is what I'm doing when I don't know what I'm doing."

--- Wernher von Braun (1912 - 1977)

The practical application of using a magnetic step function (i.e. charging a coil to establish a magnetic field, and abruptly interrupting the current flow) and then monitoring the response, was analysed in some detail by Westcott [1] and Johnson [3] in the mid 1950s.

According to Westcott, the pulse induction method provides less ambiguity, greater range and more accurate location when compared to a continuous wave method.

In the paper "A Pulsed Bomb Locator" by F.B. Johnson M.A. [3], equation 1 (at the bottom of page 2) represents the magnetic field at a buried spherical target. As is often the case in academic papers, Johnson simply says "*We thus have*" and presents the equation. So where does this equation come from?

Firstly, problems related to magnetic fields are usually quite complex, and this particular case is no exception. Consequently Johnson [3] makes use of a number of simplifying assumptions to reduce this complexity. To start with he assumes the target is a sphere which has been placed in a uniform magnetic field, with the result that the field inside the sphere is also uniform and parallel to the original direction of the applied field.

Secondly, the form of the eddy currents in the target are such that the external field produced by them (according to Lenz's Law) is a dipole field.

Thirdly, the decay time of the eddy currents is exponential - Lamb [2].

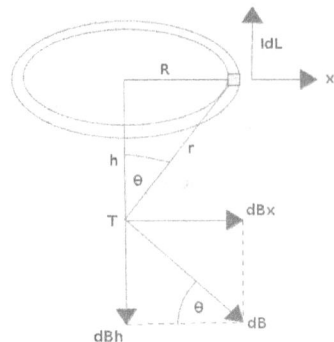

Fig. E-1: Magnetic field at a point T (target) located on the axis of a coil

Starting with the Biot-Savart law. This is an equation of the magnetic field generated by an electric current and describes the magnetic field in terms of the magnitude, direction, length and proximity of the electric current. It is named after Jean-Baptiste Biot and Felix Savart, who discovered the relationship in 1820.

$$dB = \frac{\mu_0}{4\pi} \cdot \frac{IdL}{r^2}$$
[Eq.1]

where:
μ_0 = permeability of free space ($4\pi \times 10^{-7} [N A^{-2}]$)
r = distance to the target T
dL = infinitesimal length of conductor carrying current

With reference to Fig. E-1:

From Pythagoras: $\qquad\qquad r^2 = h^2 + R^2$

and (from trigonometry): $\qquad sin\theta = \dfrac{dB_h}{dB}$

Also:

$$sin\theta = \frac{R}{\sqrt{h^2 + R^2}}$$

Substituting into Eq.1 gives:

$$\frac{dB_h}{sin\theta} = \frac{\mu_0 I dL}{4\pi} \cdot \frac{1}{(h^2 + R^2)}$$

and hence:

$$dB_h = \frac{\mu_0 I dL}{4\pi} \cdot \frac{1}{(h^2 + R^2)} \cdot \frac{R}{\sqrt{(h^2 + R^2)}}$$

Therefore we have:

$$dB_h = \frac{\mu_0 I dL}{4\pi} \cdot \frac{R}{(h^2 + R^2)^{\frac{3}{2}}} \qquad \text{[Eq.2]}$$

Equation 2 represents the magnetic field at a distance (h) from the centre of the coil, but only for an infinitesimal length of a conductor carrying an electric current I.

To get the magnetic field at a distance due to the whole coil, we need to integrate both sides of equation 2, as follows:

$$\int dB_h = \frac{\mu_0 I R}{4\pi (h^2 + R^2)^{\frac{3}{2}}} \cdot \int dL$$

Note that μ_0, I, h and R are all constants, and the integral of dL is the circumference of a circle.

Therefore we have:

$$B = \frac{\mu_0 I R}{4\pi (h^2 + R^2)^{\frac{3}{2}}} \cdot 2\pi R$$

and simplifying:

$$B = \frac{\mu_0 I}{2} \cdot \frac{R^2}{(h^2 + R^2)^{\frac{3}{2}}} \qquad \text{[Eq.3]}$$

If there are T turns on the coil, then:

$$B = \frac{\mu_0}{2}.TI.\frac{R^2}{(h^2 + R^2)^{\frac{3}{2}}} \qquad \text{[Eq.4]}$$

and (since B = μ_0 H) we have:

$$H = \frac{TI}{2}.\frac{R^2}{(h^2 + R^2)^{\frac{3}{2}}} \quad [\text{Am}^{-1}] \qquad \text{[Eq.5]}$$

where H is the magnetic field intensity, and B is the magnetic flux density (magnetic induction). Referring to Johnson's paper, Eq.5 is the same result, but given here in SI units (MKS) rather than cgs.

Johnson [3] states that, "*If the coil is critically damped, its magnetic field will decay approximately exponentially at switch-off; and it can be shown that, so long as the time constant of switch-off is less that about a tenth of the decay time of eddy currents in the object, the behaviour of the eddy currents is practically the same as if the field had been removed instantaneously, and the exact shape of the switch-off is immaterial. It is shown that under these conditions the eddy currents are such that the sphere appears to be uniformly magnetised with intensity of magnetisation I given by:*"

$$J = \frac{3H_0 e^{\frac{-t}{\tau}}}{8\pi} \; [\text{Am}^{-1}] \qquad \text{[Eq.6]}$$

where:
J = intensity of magnetization of a sphere (target object)
τ = decay time of eddy currents in the target

In Eq.6, τ is the decay time of the eddy currents, and is given by:

$$\tau = \frac{\mu a^2 \sigma}{\pi^2} \; [seconds] \qquad \text{[Eq.7]}$$

using SI units.

where:
a = radius of the sphere [m]
σ = conductivity of the material [Sm^{-1}]

Eq.7 is interesting, as it enables us to calculate the τ for various (spherical) targets, based on their radius and conductivity.

For example, using Table 1 - let's take a small 5mm diameter ball of gold. So that a = 2.5mm, $\sigma = 4.10$ x 10^7, and $\mu_0 = 4\pi$ x 10^{-7}.

$$\tau = \frac{(4\pi \times 10^{-7}) \times (2.5 \times 10^{-3})^2 \times (4.10 \times 10^7)}{\pi^2} = 32.6\mu s$$

and also a 5mm diameter ball of lead, with a conductivity $\sigma = 4.55$ x 10^6.

$$\tau = \frac{(4\pi \times 10^{-7}) \times (2.5 \times 10^{-3})^2 \times (4.55 \times 10^6)}{\pi^2} = 3.6\mu s$$

The calculated values of τ for these two targets would (at first sight) appear to contradict the fact that small gold nuggets are difficult to detect, given that the conductivity of gold is 9x the conductivity of lead. However, this is for *pure gold*, and gold found in the field has a much lower conductivity due to impurities, Plus the gold is not all in one chunk, which restricts the flow of eddy currents to a greatly reduced area, thus weakening the return signal. This would explain why gold nuggets are generally considered to be low conductivity targets, when (in fact) gold is highly conductive.

In the real world targets are rarely perfect spheres. However, it is possible (within reason) to approximate an object by defining a sphere with an identical mass, and this is exactly what Johnson [3] did in his experiments. In this case they used three cylindrical test objects of brass, duralumin[1] and mild steel, plus a 100 lb bomb. The results they obtained were in reasonable agreement, considering the number of approximations and estimates made in the calculations.

In conclusion, it appears to be possible to calculate (to a reasonable accuracy) the τ of potential targets as long as their length, breadth and height are not drastically different. This is evidenced by the tests made on the 100 lb bomb, where Johnson [3] discovered the measured signal to be 10dB higher than calculated. But this is understandable, considering how wildly the characteristics of the bomb differ from the ideal model.

Position	Name	Conductivity [Sm^{-1}]
1	Carbon (graphene)	1.00×10^8
2	Silver	6.30×10^7
3	Copper	5.96×10^7
4	Gold	4.10×10^7
5	Aluminium	3.50×10^7
6	Calcium	2.98×10^7
7	Tungsten	1.79×10^7
8	Zinc	1.69×10^7
9	Nickel	1.43×10^7
10	Lithium	1.08×10^7
11	Iron	1.00×10^7
12	Platinum	9.43×10^6
13	Tin	9.17×10^6
14	Lead	4.55×10^6
15	Titanium	2.38×10^6
16	Sea Water	4.80
17	Drinking Water	5.00×10^{-4} to 5.00×10^{-2}
18	Silicon	1.56×10^{-3}
19	Wood (damp)	10^{-4} to $\times 10^{-3}$

Table. E-1: Conductivity Chart

Notes:
1. Duralumin is a hard alloy of aluminium with copper and other elements.

References:
1. J.H. Westcott - Paper A.C. 13259 of the S.85A.C Lines and Acoustics Committee (February 1955)
2. H. Lamb M.A. - Philosophical Transactions of the Royal Society of London 174, 519-549, published 1 January 1883 "On Electrical Motions in a Spherical Conductor"

3. F.B. Johnson M.A. - A Pulsed Bomb Locator (1956)
4. https://en.wikipedia.org/wiki/Electrical_resistivity_and_conductivity

Appendix F

Sampling Integrator Analysis

"We live in a society exquisitely dependent on science and technology, in which hardly anyone knows anything about science and technology."

--- Carl Sagan (1934 - 1996)

With reference to a very interesting paper by John Alldred [1] that was passed to me by Eric Foster:

In a standard pulse induction detector, front-end signal processing usually takes the form of a high-gain broadband preamp (with high slew rate, rapid settling time and recovery from overload). From there the signal goes through a sample gate to an integrator, and then further DC amplification is applied.

The target input signal takes the form of a decaying exponential:

$$v = v_0 exp^{\frac{-t}{\tau}} \qquad \text{[Eq.1]}$$

where:

v = value of the target signal due to eddy currents induced by the transmit pulse. This cannot be observed directly due to being swamped by the reverse voltage from the coil.

v_0 = initial value at time (t=0) when the coil current is switched off.

t = elapsed time after the transmit pulse is switched off (t>0).

τ = decay time constant of the target.

At time t_1 the sample gate closes for a duration t_2. Hence t_1 represents the main sample delay, and t_2 is the main sample width. During the period when the sample gate is closed, the charge on the capacitor C is incremented by an certain amount, according to:

$$\Delta Q = \frac{1}{R_s} \int_{t_1}^{t_1 + t_2} v \; dt \qquad \text{[Eq.2]}$$

ΔQ is the incremental charge transferred to the capacitor during each sample. There is also a scale factor (R_s) which represents the resistance in series with the sample gate plus the resistance of the sample gate itself (although this is small in comparison). The capacitor charging period occurs during the time period between t_1 (gate closes) and $t_1 +$

t_2 (gate opens). Which simply means that the capacitor C is charged via the series resistor R_S when the main sample occurs, and this charge increments linearly over time in the presence of a metal target as more samples are taken.

If we substitute Eq.1 into Eq.2, we have:

$$\Delta Q = \frac{1}{R_s} \int_{t_1}^{t_1+t_2} v_0 \ \exp^{-t/\tau} \ dt$$

Then:

$$\Delta Q = \frac{v_0}{R_s} \int_{t_1}^{t_1+t_2} \exp^{-t/\tau} dt = \left[-\frac{v_0}{R_s} \frac{\exp^{-t/\tau}}{1/\tau} \right]_{t_1}^{t_1+t_2} = \left[-\frac{v_0\tau}{R_s} \exp^{-t/\tau} \right]_{t_1}^{t_1+t_2}$$

Plugging in the limits for the integral (t_1 and t_1+t_2) gives:

$$\Delta Q = \left(-\frac{v_0\tau}{R_s} \exp^{-(t_1+t_2)/\tau} \right) - \left(-\frac{v_0\tau}{R_s} \exp^{-t_1/\tau} \right)$$

$$= \left(\frac{v_0\tau}{R_s} \exp^{-t_1/\tau} \right) - \left(\frac{v_0\tau}{R_s} \exp^{-(t_1+t_2)/\tau} \right)$$

$$= \frac{v_0\tau}{R_s} \left(\exp^{-t_1/\tau} - \exp^{-(t_1+t_2)/\tau} \right)$$

Hence:

$$\Delta Q = \frac{v_0\tau \exp^{-t_1/\tau}}{R_s} \left(1 - \exp^{-t_2/\tau} \right) \qquad \text{[Eq.3]}$$

In between samples the capacitor C loses some of its charge by:

$$-\Delta Q = \frac{vT}{R_{leak}} \qquad \text{(where } R_{leak} \gg R_{S)}\text{)}$$

R_{leak} represents the leakage resistance for the charge on the capacitor.

The resultant equilibrium dc signal is:

$$v = \frac{R_{leak}}{R_s} v_0 \exp^{-t_1/\tau} \frac{\tau}{T} \left(1 - \exp^{-t_2/\tau} \right) \qquad \text{[Eq.4]}$$

Hence we arrive at an equation which describes how the received target signal is affected by the setting of t_1 (sample delay), t_2 (sample pulse width), and their relationship to τ (target decay constant).

All metal targets have an associated decay constant. This is usually represented by the Greek letter τ (tau). When a PI detector switches off its transmitter, the magnetic field of the coil collapses and causes eddy currents to flow in any nearby metal object. These eddy currents die away over time in an exponential manner, but the time taken for these currents to decay depends on the target's material, size, shape, and orientation. In general though, small gold nuggets have a very short time constant, whereas a horse shoe (for example) would have a very long time constant.

Eq.4 explains how the signal voltage at the output of the sample gate / integrator is affected when different sample delays and pulse widths are used.

As you can readily see from Eq.4, it consists of three distinct terms:

The constant term: $$v_0 \frac{R_{leak}}{R_s} \frac{\tau}{T}$$

First exponential term: $$exp^{\frac{-t_1}{\tau}}$$

Second exponential term: $$\left(1 - exp^{\frac{-t_2}{\tau}}\right)$$

For the case where $t_2 \ll \tau$, the DC equilibrium signal in Eq.4 can be approximated by:

$$v \approx v_0 \exp^{-t_1/\tau} \frac{R_{leak}}{R_s} \frac{t_2}{T}$$

We can confirm this with a simple example by proving that: $t_2 \approx \tau \left(1 - \exp^{-t_2/\tau}\right)$

Let's set $t_2 = 1\mu s$ and $\tau = 100\mu s$, so that we have a situation with a highly conductive target being detected with a very narrow sample pulse width.

Therefore: $\tau \left(1 - \exp^{-t_2/\tau}\right) = 100\mu s \left(1 - \exp^{-1/100}\right) = 0.995\mu s \approx t_2$

Hence, for the case where $t_2 \ll \tau$, we can replace both the second exponential term and τ in the constant term by t_2.

Confirming that $V \approx v_0 \exp^{-t_1/\tau} \frac{R_{leak}}{R_s} \frac{t_2}{T}$ for the case where $t_2 \ll \tau$.

Alternatively we could make $t_2 \gg \tau$, and in this instance the approximate equation becomes:

$$V = v_0 \exp^{-t_1/\tau} \frac{R_{leak}}{R_s} \frac{\tau}{T}$$

This approximation assumes that $\left(1 - \exp^{-t_2/\tau}\right)$ is close to unity, which will be true for $t_2 \gg \tau$.

Hence two basic (approximate) equations are obtained that describe the extremes of the sample pulse width settings:

$$v = v_0 exp^{-t_1/\tau} \frac{R_{leak}}{R_S} \frac{t_2}{T} \qquad \text{(where } t_2 \ll \tau) \qquad \text{[Eq.5]}$$

and:

$$v = v_0 exp^{-t_1/\tau} \frac{R_{leak}}{R_S} \frac{\tau}{T} \qquad \text{(where } t_2 \gg \tau) \qquad \text{[Eq.6]}$$

Eq.5 refers to the case where the sample pulse width is much <u>shorter</u> than the target decay constant.
Eq.6 refers to the case where the sample pulse width is much <u>longer</u> than the target decay constant.

With reference to Eq.5, and assuming that t_1 (sample delay), τ (target decay constant), and T (pulse period) are all fixed; then targets with a large τ have very little effect on v. Remember that Eq.5 is only true when t_2 is much shorter then τ. If the target changes to one with a very short decay time then t_2 may become close to, or even longer than τ, in which case this approximation will no longer be valid.

Eq.6, on the other hand, is only valid for the condition where t_2 is much longer than τ. In this case, v is much less affected by the choice of sample pulse width for targets with a small decay constant.

According to Ref-1, since $R_{leak} \gg R_S$ and $T \gg t_2$, the dimensionless scale factor $\dfrac{R_{leak}}{R_S} \dfrac{\tau}{T}$ is (in practice) typically in the region of unity, and can therefore be excluded from the equation.

For example:

$$\frac{R_{leak}}{R_s} \frac{\tau}{T} = \frac{100k}{1k} \frac{10u}{1m} = 1$$

Hence we have:

$$v = v_0 exp^{-t_1/\tau} \left(1 - exp^{-t_2/\tau}\right) \qquad \text{[Eq.7]}$$

At this point it would be interesting to produce a graphical representation of (Eq.7) using the LTspice simulator (see Fig. F-1).

The simulation is set up so that the parameter tau is replaced by the simulation variable *time*, which means that the x-axis of the time domain simulation represents τ, and therefore the plot window (Fig. F-2) shows how the sampling integrator output (*v*) changes for various values of target time constant.

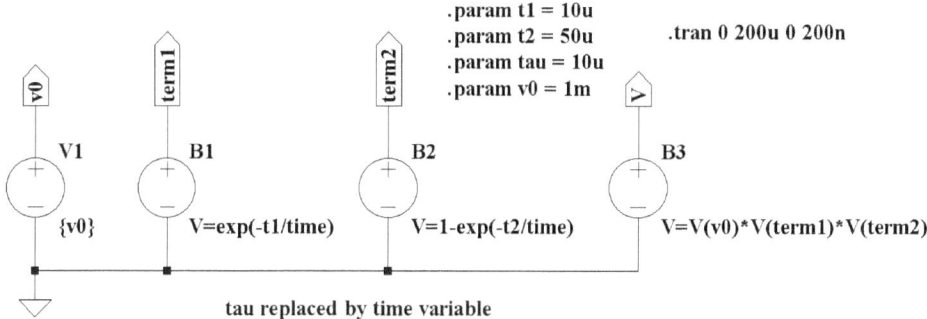

tau replaced by time variable

Fig. F-1: LTspice simulation of Eq.7 using time to replace tau.

The simulation uses behavioural voltage sources (B elements) to define the individual terms of Eq.7, which enables each of the terms to be plotted separately if desired. The B element (B3) calculates the final result. (See Fig. F-2.) The integrator output voltage (*v*) represents that part of the input signal attributable to the target.

Note that in Fig. F-1, the sample delay (t_1) is held constant at 10μs, and the sample pulse width (t_2) at 50μs. The default value of the target decay constant (τ) is set to 10μs (the approximate decay constant for a U.S. Nickel) but this is effectively overridden by the *time* variable, so that we can sweep the value of τ from 0μs to 200μs and readily see its effect on the integrator output (*v*).

When the target τ is zero, *v* is quite obviously going to be zero. As the τ increases, *v* becomes larger in amplitude before peaking when τ reaches 29μs, and then decreasing exponentially as τ is increased further. Clearly a sample delay of 10μs, together with a sample pulse width of 50μs will therefore be more sensitive to targets with a decay constant of around 29μs.

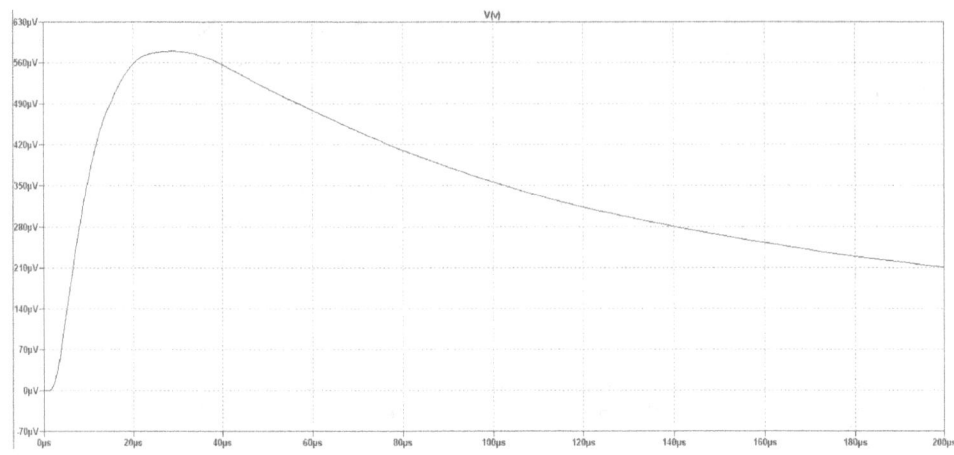

Fig. F-2: LTspice simulation results for Eq.7 (v versus τ)

If we perform a parametric sweep of the sample delay (t_1) using:

.step param t1 0 50u 10u

we can also determine how the output voltage (v) is affected by variations in both target τ and sample delay (t_1).

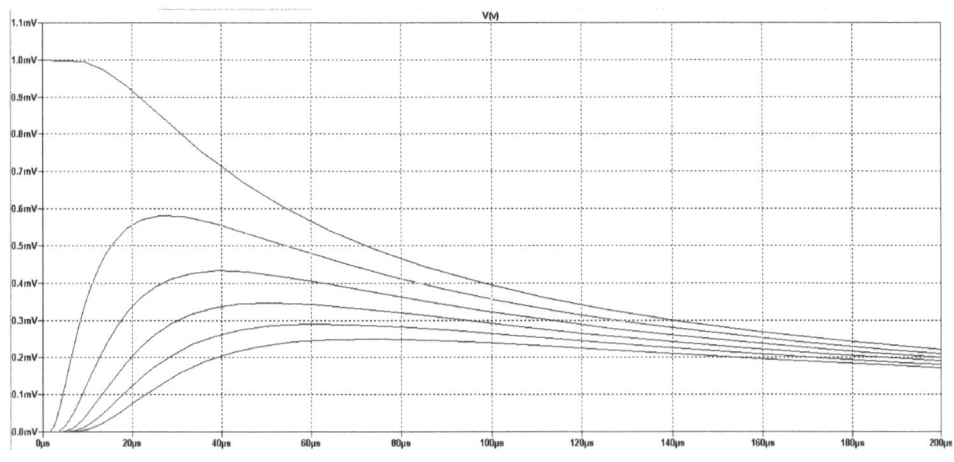

Fig. F-3: LTspice simulation results for Eq.7 (v versus τ) with sweep of t_1

In Fig. F-3 the parametric sweep results for t_1 start with 0us at the top, down to 50us at the bottom. From these waveforms it is evident that increasing the sample delay will produce a reduction in amplitude as a result of eddy current decay, as would be expected. It is interesting to note how this also affects the sensitivity to τ.

In the same way as we replaced the target time constant (τ) with the *time* variable, we can repeat this for the sample delay (t_1). This is shown in the schematic in Fig. F-4.

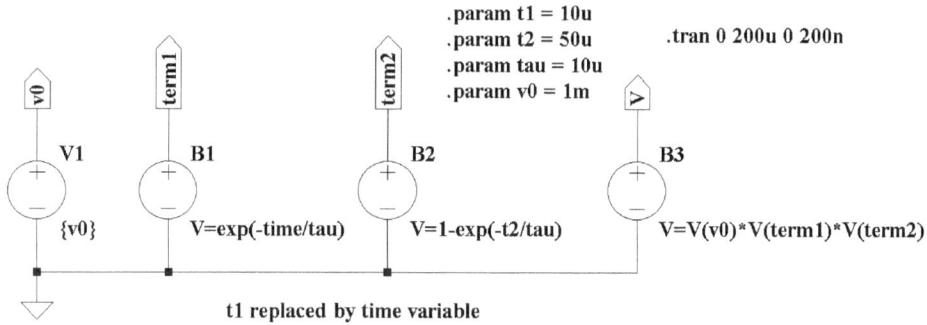

t1 replaced by time variable

Fig. F-4: LTspice simulation of Eq.7 using time to replace t_1.

In Fig. F-4, the target time constant (τ) is held constant at 10μs and the sample pulse width (t_2) at 50us. The default value of the sample delay (t_1) is set to 10us, but this is effectively overridden by the *time* variable, so that we can sweep the value of t_1 from 0us to 50us and determine its effect on the integrator output (v).

The plot results in Fig. F-5 are exactly as you would expect. That is, the output voltage (v) decreases in amplitude as the sample delay is increased. But what happens if we also sweep the value of τ together with the sample delay (t_1)?

Again, we can easily do this using a parametric sweep:

.step param tau 10u 50u 10u

The plot results are shown in Fig. F-6, where the fastest decay curve is associated with a τ of 10us, and the slowest with 50us. Note that a sweep with τ equal to zero was not included, as this obviously produces a zero output.

Finally, we can perform similar simulations by replacing the sample pulse width (t_2) with the *time* variable. The schematic is shown in Fig. F-7, and the plot results are in Fig. F-8.

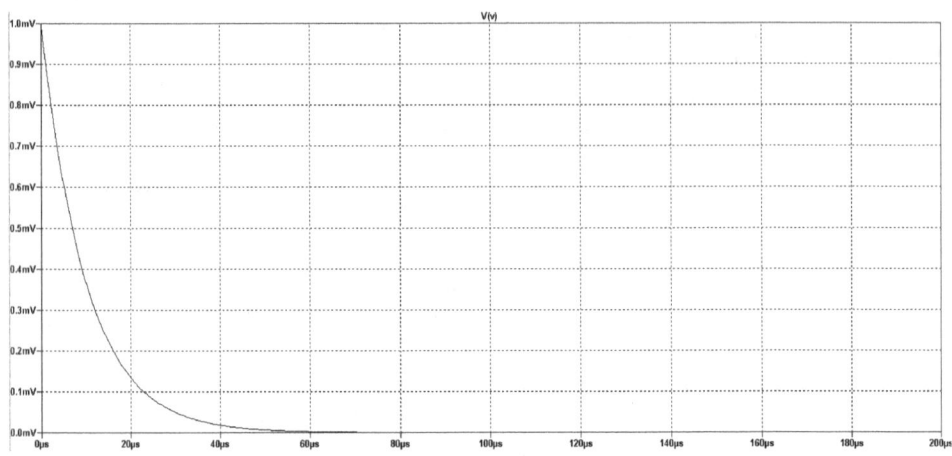

Fig. F-5: LTspice simulation results for Eq.7 (v versus t_1)

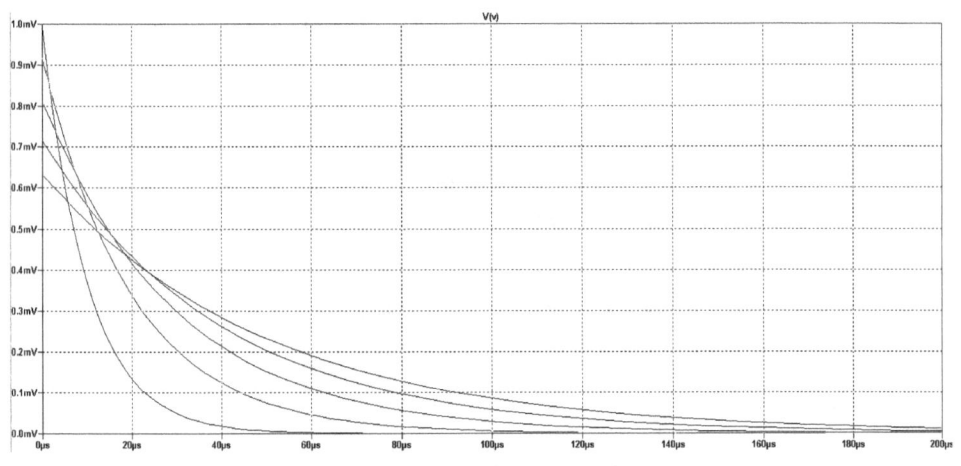

Fig. F-6: LTspice simulation results for Eq.7 (v versus t_1) with sweep of τ

The Voodoo Project

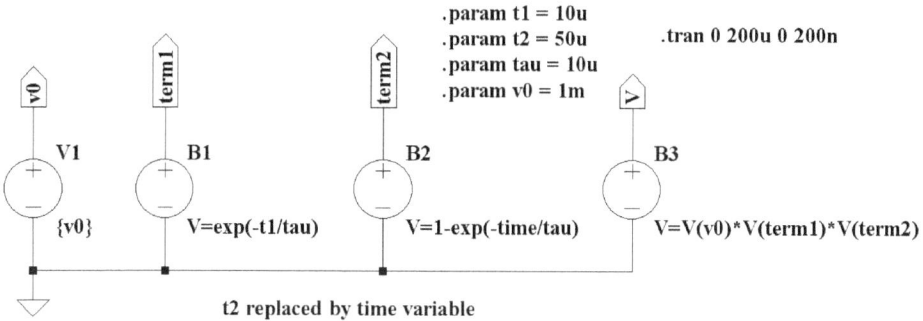

.param t1 = 10u
.param t2 = 50u
.param tau = 10u
.param v0 = 1m

.tran 0 200u 0 200n

V1
{v0}

B1
V=exp(-t1/tau)

B2
V=1-exp(-time/tau)

B3
V=V(v0)*V(term1)*V(term2)

t2 replaced by time variable

Fig. F-7: LTspice simulation of Eq.7 using time to replace t_2.

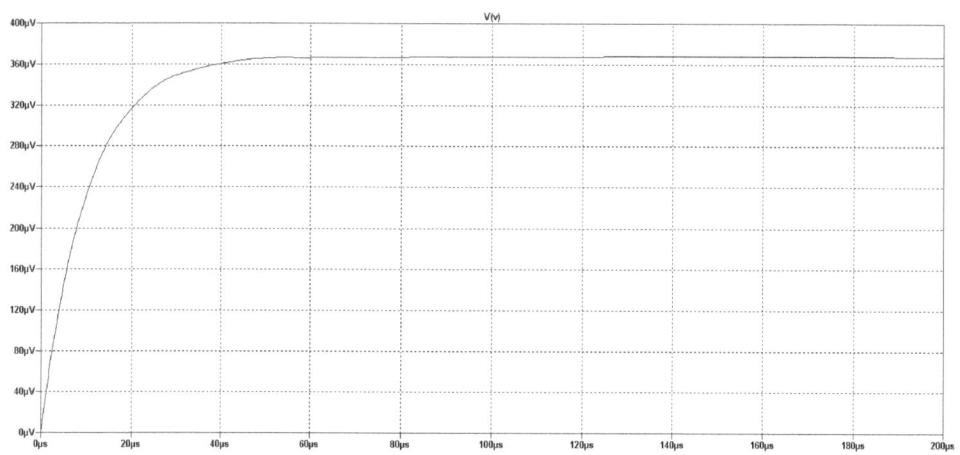

Fig. F-8: LTspice simulation results for Eq.7 (v versus t_2)

It is interesting to note in Fig. F-8 that increasing the sample pulse width (t_2) beyond 50us provides no further increase in the amplitude of the output signal (v) when the target time constant (τ) is fixed at 10us. Is this true for other values of target time constant (τ)? Yet again, we can examine the problem by adding a parametric sweep, as follows:

.step param tau 10u 50u 10u

The answer is revealed in Fig. F-9.

Sampling Integrator Analysis   **123**

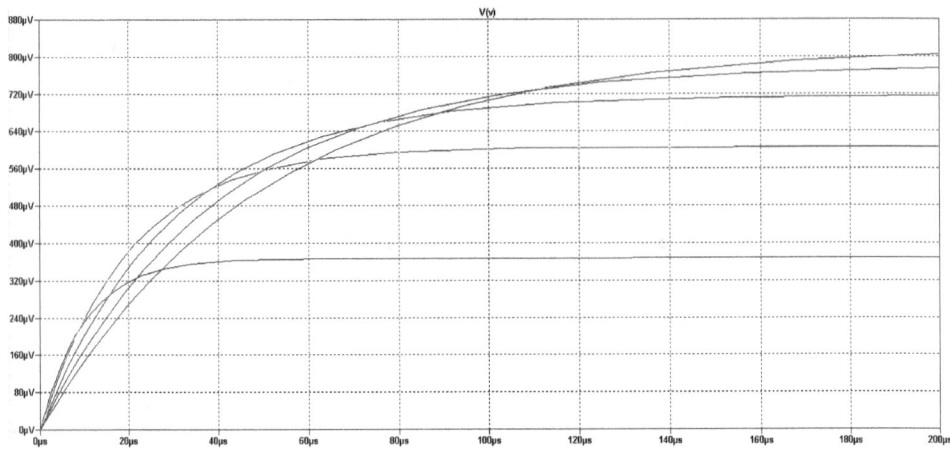

Fig. F-9: LTspice simulation results for Eq.7 (v versus t_2) with sweep of τ

The lower curve in Fig. F-9 is associated with the target τ of 10us, and you can see how increasing values of τ level off at longer sample delays.

With simple integrator circuits the baseline of the signal can deviate significantly from zero, and also fluctuate due to noise. If the baseline offset voltage is v_n, then we have (at the integrator output):

$$\frac{R_{leak}}{R_s}\frac{1}{T}\int_{t_1}^{t_1+t_2} v_n dt = \frac{R_{leak}}{R_s}\frac{t_2}{T}v_n$$

As can be seen from the various simulation results, it would appear that t_2 should be made as wide as possible, but a side effect of this is to degrade the signal-to-noise ratio.

Another problem is detection of the Earth's magnetic field when the coil is in motion. Even though the Earth's magnetic field is very weak, the high gain of a PI detector can produce an audio response which is evidenced by a beep at the end of each swing.

To provide Earth-field elimination (EFE) we need to add a second sample gate which is opened at a time (which is less than the TX period T) and remains open for the same length of time as t_2. The idea is that any signal from the Earth-field (EF) changes very slowly in comparison to the target signal, and remains substantially unchanged during the TX period. Since the EF is present in both samples and the target signal has reduced significantly by the time the later t_3 sample occurs, then subtracting the t_3 sample from the t_2 sample will provide EFE.

The signal from the t_3 (EF) sample will be smaller than the t_2 sample by a factor $exp^{-t_3/\tau}$, but this will be insignificant compared to that from t_2 and (as long as t_3 is positioned towards the end of the TX period, away from t_2) there will be virtually no reduction in signal amplitude due to the EFE process.

Of course, the values of t_1, t_2 and t_3 can be optimised for a particular target τ, but this will also weaken the response to targets outside this value. It would appear that one possible solution may be to provide two (or more) receiver channels that operate with different values of t_1, t_2 and t_3 that are optimised to a different target τ. In this way it might be possible to accommodate a wider range of target responses, and provide an indication of approximate τ to give an idea of target size.

References:
1. John C. Alldred MA MSc (1941 - 2002) - Protovale Oxford Limited - technical paper supplied by Eric Foster - section: Electronic Signal Processing

Embedded Software – Full Source Code

"Any fool can write code that a computer can understand. Good programmers write code that humans can understand."

--- Martin Fowler – British software developer, author and international public speaker

```c
/*
 * Project name:
 *    Voodoo - PI/VLF Hybrid Metal Detector
 * Copyright:
 *    (c) George Overton, 2020.
 * Revision History:
 *    - V1.0 Initial release
 * Description:
 *    Pulse Induction / Continuous Wave Hybrid Metal Detector
 * Test configuration:
 *    MCU:            PIC18F4520
 *    Oscillator:     20 MHz Crystal
 *    Ext. Modules:   None.
 *    SW:             mikroC PRO for PIC
 *                    http://www.mikroe.com/mikroc/pic/
 * NOTES:
 *    Sync signal must be present for +5V supply to work.
 */

// Assign port I/O macro definitions
#define TRISA_INIT 0x0F;
#define TRISB_INIT 0xC0;
#define TRISC_INIT 0xF0;
#define TRISD_INIT 0x38;
#define TRISE_INIT 0x06;

// LCD assignments
sbit LCD_D4 at RB3_bit;                     // Data line 4
sbit LCD_D5 at RB2_bit;                     // Data line 5
sbit LCD_D6 at RB1_bit;                     // Data line 6
sbit LCD_D7 at RB0_bit;                     // Data line 7
sbit LCD_RS at RB5_bit;                     // Register select
sbit LCD_EN at RB4_bit;                     // Enable

// LCD Port direction
sbit LCD_RS_Direction at TRISB5_bit;        // Register select
sbit LCD_EN_Direction at TRISB4_bit;        // Enable
sbit LCD_D7_Direction at TRISB0_bit;        // Data bit 7
sbit LCD_D6_Direction at TRISB1_bit;        // Data bit 6
sbit LCD_D5_Direction at TRISB2_bit;        // Data bit 5
sbit LCD_D4_Direction at TRISB3_bit;        // Data bit 4

// Port assignments
sbit debug at LATD2_bit;                    // Debug output
sbit EN1 at LATD6_bit;                      // Enable MOSFET
sbit ps_sync at LATC0_bit;                  // Power supply sync pulse
sbit main_pulse at LATC3_bit;               // Main sample pulse
sbit efe_pulse at LATD0_bit;                // EFE sample pulse
sbit disc_pulse at LATC1_bit;               // DISC_sample pulse
sbit audio_en at LATA4_bit;                 // Audio enable
```

```c
sbit menu_btn_port at PORTC.B6;           // Keypad menu button
sbit up_btn_port at PORTC.B5;             // Keypad up button
sbit down_btn_port at PORTC.B4;           // Keypad down button
sbit enter_btn_port at PORTD.B3;          // Keypad enter button

// General macro definitions
#define EN1_on 1;                          // Turn on MOSFET
#define EN1_off 0;                         // Turn off MOSFET
#define main_on 0;                         // Turn on main sample
#define main_off 1;                        // Turn off main sample
#define efe_on 0;                          // Turn on EFE sample
#define efe_off 1;                         // Turn off EFE sample
#define disc_on 0;                         // Turn on DISC sample
#define disc_off 1;                        // Turn off DISC sample
#define audio_on 1;                        // Audio enabled
#define audio_off 0;                       // Audio disabled

// General Constants
const debounce = 100;                      // 100ms debounce time
const btn_off = 1;                         // Keypad button off
const btn_on = 0;                          // Keypad button on
const menu_btn = 1;                        // Keypad menu button
const up_btn = 2;                          // Keypad up button
const down_btn = 3;                        // Keypad down button
const enter_btn = 4;                       // Keypad enter button
const menu_active = 1;                     // Indicates menu is in use
const menu_inactive = 0;                   // Indicates menu is not being used
const pulse = 0;                           // Indicates pulse detection mode
const hybrid = 1;                          // Indicates hybrid detection mode
const blk_limit = 7;                       // Block display limit

// Timer offset constants (us)
const txon_offset = 4;
const txpd_offset = 8;
const main_dly_offset = 5;
const main_smpl_offset = 4;
const efe_dly_offset = 10;
const disc_dly_offset = 9;
const disc_smpl_offset = 6;

// Variables
unsigned i;                                // Generic variable
unsigned tmp;                              // Temporary variable
char int_state;                            // State machine for interrupt routine
char main_state;                           // State machine for main routine
char hybrid_cycle;                         // Hybrid operating cycle (0 = pulse, 1 = disc)
unsigned t;                                // Temporary storage for ADC readings
unsigned vbatt;                            // Battery voltage
unsigned pi_target;                        // PI target voltage
unsigned disc_target;                      // DISC target voltage
char loop_count;                           // Main loop count
unsigned pi_array[64];                     // Running average array for PI target voltages
char pi_pointer;                           // Pointer into PI array
unsigned pi_tmp;                           // Temporary variable for PI array calculations
unsigned pi_total;                         // Total of pi_array contents
unsigned disc_array[64];                   // Running average array for disc target voltages
char disc_pointer;                         // Pointer into disc array
unsigned disc_tmp;                         // Temporary variable for disc
unsigned disc_total;                       // Total of disc_array contents
unsigned short pi_adc = 0;                 // PI channel ADC
unsigned short disc_adc = 1;               // Disc channel ADC
unsigned short batt_adc = 7;               // Battery monitor ADC
char keypad_btn;                           // Keypad button that has been pressed
char result;                               // Records result of a button push
char menu_flag;                            // Flags state of menu system (active or inactive)
char menu_disp;                            // Indicates current menu display screen
char detect_mode;                          // Operating mode of detector (0 = pulse, else =
hybrid)                                    hybrid)
unsigned pi_thr;                           // Audio threshold for PI channel
unsigned disc_thr;                         // Audio threshold for disc channel
char thr_disp;                             // Audio threshold for display purposes
char txon;                                 // TX pulse width (us)
```

```c
char txonh;                // TX pulse width high byte
char txonl;                // TX pulse width low byte
unsigned txpd;             // TX pulse period (us)
char txpdh;                // TX pulse period high byte
char txpdl;                // TX pulse period low byte
char main_dly;             // Main sample delay (us)
char main_dlyh;            // Main sample delay high byte
char main_dlyl;            // Main sample delay low byte
char main_smpl;            // Main sample delay pulse width (us)
char main_smplh;           // Main sample pulse width high byte
char main_smpll;           // Main sample pulse width low byte
unsigned efe_dly;          // EFE sample delay (us)
char efe_dlyh;             // EFE sample delay high byte
char efe_dlyl;             // EFE sample delay low byte
char efe_smplh;            // EFE sample pulse width high byte
char efe_smpll;            // EFE sample pulse width low byte
char pi_num;               // Number of PI readings to average
char disc_dly;             // DISC sample delay (us)
char disc_dlyh;            // DISC sample delay high byte
char disc_dlyl;            // DISC sample delay low byte
char disc_smpl;            // DISC sample pulse width (us)
char disc_smplh;           // DISC sample pulse width high byte
char disc_smpll;           // DISC sample pulse width low byte
char disc_num;             // Number of disc readings to average
char accept;               // Accept (non-ferrous) counter
char accept_blk;           // Accept block display counter
char reject;               // Reject (ferrous)counter
char reject_blk;           // Reject block display counter
char update_disp;          // Update display flag
char meter_zero;           // Meter zero counter
char meter_zero_limit;     // Meter zero counter limit
char counter_limit;        // Accept and reject counter limit

// Interrupt routine
void interrupt() {
  switch (int_state) {
    case 0:                           // TX on - charge coil
      EN1 = EN1_on;                   // Turn on MOSFET
      TMR0H = txonh;                  // Load TMR0 for TX pulse
      TMR0L = txonl;
      TMR1H = txpdh;                  // Load TMR1 for TX period
      TMR1L = txpdl;
      ps_sync = 1;                    // Sync power supply
      debug = 1;
      PIR1 = 0x00;                    // Clear TMR1 overflow interrupt flag
      INTCON = 0xE0;                  // Clear TMR0 overflow interrupt flag
      if (hybrid_cycle == 0) {
        int_state = 1;                // Pulse sample
      } else {
        int_state = 6;                // Disc sample
      }
      break;
    case 1:                           // TX off - discharge coil
      EN1 = EN1_off;                  // Turn off MOSFET
      TMR0H = main_dlyh;              // Load TMR0 for main sample delay
      TMR0L = main_dlyl;
      ps_sync = 0;                    // Sync power supply
      debug = 0;
      INTCON = 0xE0;                  // Clear TMR0 overflow interrupt flag
      int_state = 2;
      break;
    case 2:                           // Main sample pulse on
      main_pulse = main_on;           // Turn on main sample
      TMR0H = main_smplh;             // Load TMR0 for main sample width
      TMR0L = main_smpll;
      ps_sync = 1;                    // Sync power supply
      INTCON = 0xE0;                  // Clear TMR0 overflow interrupt flag
      int_state = 3;
      break;
    case 3:                           // Main sample pulse off
      main_pulse = main_off;          // Turn off main sample
      TMR0H = efe_dlyh;               // Load TMR0 for EFE sample delay
```

```
        TMR0L = efe_dlyl;
        ps_sync = 0;                                // Sync power supply
        INTCON = 0xE0;                              // Clear TMR0 overflow interrupt flag
        int_state = 4;
        break;
      case 4:                                       // EFE sample pulse on
        efe_pulse = efe_on;                         // Turn on EFE sample
        TMR0H = efe_smplh;                          // Load TMR0 for EFE sample width
        TMR0L = efe_smpll;
        ps_sync = 1;                                // Sync power supply
        INTCON = 0xE0;                              // Clear TMR0 overflow interrupt flag
        int_state = 5;
        break;
      case 5:                                       // EFE sample pulse off
        efe_pulse = efe_off;                        // Turn off EFE sample
        TMR0H = 0x00;                               // Load TMR0 with maximum delay
        TMR0L = 0x00;
        ps_sync = 0;
        INTCON = 0xE0;                              // Clear TMR0 overflow interrupt flag
        if (detect_mode == pulse) {                 // Check detector mode (pulse or hybrid)
          hybrid_cycle = 0;                         // Continue with pulse sampling
        } else {
          hybrid_cycle = 1;                         // Switch to disc sampling
        }
        int_state = 0;
        break;
      case 6:                                       // TX off - discharge coil
        EN1 = EN1_off;                              // Turn off MOSFET
        TMR0H = disc_dlyh;                          // Load TMR0 for DISC sample delay
        TMR0L = disc_dlyl;
        ps_sync = 0;                                // Sync power supply
        debug = 0;
        INTCON = 0xE0;                              // Clear TMR0 overflow interrupt flag
        int_state = 7;
        break;
      case 7:                                       // Disc sample pulse on
        disc_pulse = disc_on;                       // Turn on disc sample pulse
        TMR0H = disc_smplh;                         // Load TMR0 for disc sample pulse width
        TMR0L = disc_smpll;
        ps_sync = 1;                                // Sync power supply
        INTCON = 0xE0;                              // Clear TMR0 overflow interrupt flag
        int_state = 8;
        break;
      case 8:                                       // Disc sample pulse off
        disc_pulse = disc_off;                      // Turn off disc sample pulse
        TMR0H = 0x00;                               // Load TMR0 with maximum delay
        TMR0L = 0x00;
        ps_sync = 0;                                // Sync power supply
        INTCON = 0xE0;                              // Clear TMR0 overflow interrupt flag
        hybrid_cycle = 0;                           // Switch to pulse sampling
        int_state = 0;
        break;
  }
}

// Extract requested digit from value and display at selected row and column on LCD
void extract_and_disp(
  unsigned value,                                   // Number to display on LCD
  unsigned divisor,                                 // Selects which digit to extract
  char row,                                         // Select row on LCD
  char col) {                                       // Select column on LCD
  char ch;                                          // Digit to display
  ch = value / divisor % 10;                        // Extract selected digit
  Lcd_Chr(row, col, 48 + ch);                       // Display selected digit
}

// Acquire battery voltage
void battery_measure() {
  t = 0;                                            // Clear battery voltage accumulator
  for (i = 1; i <= 8; i++) {
    t += ADC_Read(batt_adc);                        // Accumulate battery voltage readings
  }
```

```
    t = t / 8;                              // Average readings
    t = (5000 - (t * 5)) * 3 / 100;         // Calculate battery voltage
}

// Show battery voltage on LCD
void battery_display() {
    extract_and_disp(t, 100, 1, 10);
    extract_and_disp(t, 10, 1, 11);
    Lcd_Out(1, 12, ".");
    extract_and_disp(t, 1, 1, 13);
    Lcd_Out(1, 14, "V ");
}

// Display appropriate battery symbol
void battery_symbol() {
    if (t >= 120)
        Lcd_Chr(1, 16, 0);
    else if (t >= 117)
        Lcd_Chr(1, 16, 1);
    else if (t >= 114)
        Lcd_Chr(1, 16, 2);
    else if (t >= 111)
        Lcd_Chr(1, 16, 3);
    else if (t >= 108)
        Lcd_Chr(1, 16, 4);
    else if (t >= 105)
        Lcd_Chr(1, 16, 5);
    else
        Lcd_Out(1, 16, "X");
}

void calc_txon() {                          // Calculate TX pulse width
    tmp = 65535 - ((txon - txon_offset) / 2) * 10;
    txonh = tmp >> 8;
    txonl = tmp - (txonh << 8);
}

void calc_txpd() {                          // Calculate TX period
    tmp = 65535 - ((txpd - txpd_offset) / 2) * 10;
    txpdh = tmp >> 8;
    txpdl = tmp - (txpdh << 8);
}

void calc_main_dly() {                      // Calculate main sample delay
    tmp = 65535 - ((main_dly - main_dly_offset) / 2) * 10;
    main_dlyh = tmp >> 8;
    main_dlyl = tmp - (main_dlyh << 8);
}

void calc_main_smpl() {                     // Calculate main sample pulse width
    tmp = 65535 - ((main_smpl - main_smpl_offset) / 2) * 10;
    main_smplh = tmp >> 8;
    main_smpll = tmp - (main_smplh << 8);
}

void calc_efe_dly() {                       // Calculate EFE sample delay
    tmp = 65535 - ((efe_dly - efe_dly_offset) / 2) * 10;
    efe_dlyh = tmp >> 8;
    efe_dlyl = tmp - (efe_dlyh << 8);
}

void calc_efe_smpl() {                      // Calculate EFE sample pulse width
    efe_smplh = main_smplh;                 // EFE sample pulse width must equal main sample
    efe_smpll = main_smpll;
}

void calc_disc_dly() {                      // Calculate disc sample delay
    tmp = 65535 - ((disc_dly - disc_dly_offset) / 2) * 10;
    disc_dlyh = tmp >> 8;
    disc_dlyl = tmp - (disc_dlyh << 8);
}
```

```
void calc_disc_smpl() {                          // Calculate disc sample pulse width
  tmp = 65535 - ((disc_smpl - disc_smpl_offset) / 2) * 10;
  disc_smplh = tmp >> 8;
  disc_smpll = tmp - (disc_smplh << 8);
}

void clear_arrays() {                            // Clear both PI and disc arrays
  for (i = 0; i <= (pi_num - 1); i++) {
    pi_array[i] = 0;                             // Clear PI array
    disc_array[i] = 0;                           // Clear disc array
  }
  pi_pointer = 0;                                // Reset PI array pointer
  pi_total = 0;                                  // Clear PI array total
  disc_pointer = 0;                              // Reset disc array pointer
  disc_total = 0;                                // Clear disc array total;
}

void write_eeprom() {
  EEPROM_Write(0, 0xAA);                         // Initialize EEPROM
  EEPROM_Write(1, detect_mode);                  // Write detection mode
  tmp = pi_thr >> 8;                             // Extract PI threshold high byte
  EEPROM_Write(2, tmp);                          // Write PI threshold high byte
  tmp = (pi_thr << 8) >> 8;                      // Extract PI threshold low byte
  EEPROM_Write(3, tmp);                          // Write PI threshold low byte
  EEPROM_Write(4, txon);                         // Write TX pulse width
  tmp = txpd >> 8;                               // Extract TX period high byte
  EEPROM_Write(5, tmp);                          // Write TX period high byte
  tmp = (txpd << 8) >> 8;                        // Extract TX period low byte
  EEPROM_Write(6, tmp);                          // Write TX period low byte
  EEPROM_Write(7, main_dly);                     // Write main sample delay
  EEPROM_Write(8, main_smpl);                    // Write main sample pulse width
  tmp = efe_dly >> 8;                            // Extract EFE sample delay high byte
  EEPROM_Write(9, tmp);                          // Write EFE sample delay high byte
  tmp = (efe_dly << 8) >> 8;                     // Extract EFE sample delay low byte
  EEPROM_Write(10, tmp);                         // Write EFE sample delay low byte
  EEPROM_Write(11, pi_num);                      // Write number of PI readings to average
  tmp = disc_thr >> 8;                           // Extract disc threshold high byte
  EEPROM_Write(12, tmp);                         // Write disc threshold high byte
  tmp = (disc_thr << 8) >> 8;                    // Extract disc threshold low byte
  EEPROM_Write(13, tmp);                         // Write disc threshold low byte
  EEPROM_Write(14, disc_dly);                    // Write disc sample delay
  EEPROM_Write(15, disc_smpl);                   // Write disc sample pulse width
  EEPROM_Write(16, disc_num);                    // Write number of disc readings to average
  EEPROM_Write(17, meter_zero_limit);            // Write meter zero limit
  EEPROM_Write(18, counter_limit);               // Write accept and reject counter limit
}

// Scan keypad for a key press
char scan_keypad() {
  result = 0;
  if (menu_btn_port == btn_on) {                 // Check menu button
    result = menu_btn;
    Delay_ms(debounce);
    while (menu_btn_port == btn_on) {}
    Delay_ms(debounce);
  } else {
    if (up_btn_port == btn_on) {                 // Check up button
      result = up_btn;
      Delay_ms(debounce);
      while (up_btn_port == btn_on) {}
      Delay_ms(debounce);
    } else {
      if (down_btn_port == btn_on) {             // Check down button
        result = down_btn;
        Delay_ms(debounce);
        while (down_btn_port == btn_on) {}
        Delay_ms(debounce);
      } else {
        if (enter_btn_port == btn_on) {          // Check enter button
          result = enter_btn;
          Delay_ms(debounce);
          while (enter_btn_port == btn_on) {}
```

```
          Delay_ms(debounce);
        }
      }
    }
  }
  return result;
}

// Menu system
void menu_system() {
  PIE1 = 0x00;                              // Disable TMR1 interrupt
  INTCON = 0x00;                            // Disable global, peripheral and TMR0 interrupts
  menu_flag = menu_active;                  // Flag menu system as active
  EN1 = EN1_off;                            // Turn off MOSFET to save power
  audio_en = audio_off;                     // Disable audio while in menu system
  menu_disp = 0;                            // Set menu display to first screen
  while (menu_flag == menu_active) {        // Enter menu system
    switch (menu_disp) {
    case 0:
      Lcd_Out(1, 1, " DETECTING MODE ");
      if (detect_mode == pulse) {
        Lcd_Out(2, 1, "     PULSE        ");
      } else {
        Lcd_Out(2, 1, "      HYBRID      ");
      }
      if ((keypad_btn == up_btn) || (keypad_btn == down_btn)) { // Toggle detection mode
        if (detect_mode == pulse) {
          detect_mode = hybrid;
        } else {
          detect_mode = pulse;
        }
      }
      break;
    case 1:
      Lcd_Out(1, 1, "  PI THRESHOLD  ");
      Lcd_Out(2, 1, "          ");
      thr_disp = pi_thr - 511;
      extract_and_disp(thr_disp, 10, 2, 8);
      extract_and_disp(thr_disp, 1, 2, 9);
      Lcd_Out(2, 10, "        ");
      if (keypad_btn == up_btn) {
        pi_thr++;
        if (pi_thr > 531) {pi_thr = 531;}
      }
      if (keypad_btn == down_btn) {
        pi_thr--;
        if (pi_thr < 511) {pi_thr = 511;}
      }
      break;
    case 2:
      Lcd_Out(1, 1, " PI PULSE WIDTH ");
      Lcd_Out(2, 1, "       ");
      extract_and_disp(txon, 100, 2, 6);
      extract_and_disp(txon, 10, 2, 7);
      extract_and_disp(txon, 1, 2, 8);
      Lcd_Out(2, 9, "us        ");
      if (keypad_btn == up_btn) {
        txon += 10;
        if (txon > 180) {txon = 180;}
      }
      if (keypad_btn == down_btn) {
        txon -= 10;
        if (txon < 10) {txon = 10;}
      }
      break;
    case 3:
      Lcd_Out(1, 1, "PI SAMPLE DELAY ");
      Lcd_Out(2, 1, "        ");
      extract_and_disp(main_dly, 10, 2, 7);
      extract_and_disp(main_dly, 1, 2, 8);
      Lcd_Out(2, 9, "us        ");
      if (keypad_btn == up_btn) {
```

```
      main_dly++;
      if (main_dly > 50) {main_dly = 50;}
    }
    if (keypad_btn == down_btn) {
      main_dly--;
      if (main_dly < 15) {main_dly = 15;}
    }
    break;
  case 4:
    Lcd_Out(1, 1, "PI SAMPLE WIDTH ");
    Lcd_Out(2, 1, "         ");
    extract_and_disp(main_smpl, 100, 2, 6);
    extract_and_disp(main_smpl, 10, 2, 7);
    extract_and_disp(main_smpl, 1, 2, 8);
    Lcd_Out(2, 9, "us       ");
    if (keypad_btn == up_btn) {
      main_smpl++;
      if (main_smpl > 60) {main_smpl = 60;}
    }
    if (keypad_btn == down_btn) {
      main_smpl--;
      if (main_smpl < 20) {main_smpl = 20;}
    }
    break;
  case 5:
    Lcd_Out(1, 1, "   PI AVERAGE   ");
    Lcd_Out(2, 1, "          ");
    extract_and_disp(pi_num, 10, 2, 8);
    extract_and_disp(pi_num, 1, 2, 9);
    Lcd_Out(2, 10, "         ");
    if (keypad_btn == up_btn) {
      pi_num++;
      if (pi_num > 64) {pi_num = 64;}
    }
    if (keypad_btn == down_btn) {
      pi_num--;
      if (pi_num < 1) {pi_num = 1;}
    }
    break;
  case 6:
    Lcd_Out(1, 1, " DISC THRESHOLD ");
    Lcd_Out(2, 1, "         ");
    thr_disp = disc_thr - 511;
    extract_and_disp(thr_disp, 10, 2, 8);
    extract_and_disp(thr_disp, 1, 2, 9);
    Lcd_Out(2, 10, "         ");
    if (keypad_btn == up_btn) {
      disc_thr++;
      if (disc_thr > 531) {disc_thr = 531;}
    }
    if (keypad_btn == down_btn) {
      disc_thr--;
      if (disc_thr < 511) {disc_thr = 511;}
    }
    break;
  case 7:
    Lcd_Out(1, 1, "DISC SMPL DELAY ");
    Lcd_Out(2, 1, "        ");
    extract_and_disp(disc_dly, 100, 2, 6);
    extract_and_disp(disc_dly, 10, 2, 7);
    extract_and_disp(disc_dly, 1, 2, 8);
    Lcd_Out(2, 9, "us       ");
    if (keypad_btn == up_btn) {
      disc_dly++;
      if (disc_dly > 100) {disc_dly = 100;}
    }
    if (keypad_btn == down_btn) {
      disc_dly--;
      if (disc_dly < 20) {disc_dly = 20;}
    }
    break;
  case 8:
```

```
        Lcd_Out(1, 1, "DISC SMPL WIDTH ");
        Lcd_Out(2, 1, "        ");
        extract_and_disp(disc_smpl, 100, 2, 6);
        extract_and_disp(disc_smpl, 10, 2, 7);
        extract_and_disp(disc_smpl, 1, 2, 8);
        Lcd_Out(2, 9, "us      ");
        if (keypad_btn == up_btn) {
          disc_smpl++;
          if (disc_smpl > 100) {main_smpl = 100;}
        }
        if (keypad_btn == down_btn) {
          disc_smpl--;
          if (disc_smpl < 10) {disc_smpl = 10;}
        }
        break;
      case 9:
        Lcd_Out(1, 1, "  DISC AVERAGE  ");
        Lcd_Out(2, 1, "        ");
        extract_and_disp(disc_num, 10, 2, 8);
        extract_and_disp(disc_num, 1, 2, 9);
        Lcd_Out(2, 10, "       ");
        if (keypad_btn == up_btn) {
          disc_num++;
          if (disc_num > 64) {disc_num = 64;}
        }
        if (keypad_btn == down_btn) {
          disc_num--;
          if (disc_num < 1) {disc_num = 1;}
        }
        break;
      case 10:
        Lcd_Out(1, 1, "METER ZERO LIMIT");
        Lcd_Out(2, 1, "        ");
        extract_and_disp(meter_zero_limit, 10, 2, 8);
        extract_and_disp(meter_zero_limit, 1, 2, 9);
        Lcd_Out(2, 10, "       ");
        if (keypad_btn == up_btn) {
          meter_zero_limit++;
          if (meter_zero_limit > 20) {meter_zero_limit = 20;}
        }
        if (keypad_btn == down_btn) {
          meter_zero_limit--;
          if (meter_zero_limit < 5) {meter_zero_limit = 5;}
        }
        break;
      case 11:
        Lcd_Out(1, 1, " COUNTER LIMIT  ");
        Lcd_Out(2, 1, "        ");
        extract_and_disp(counter_limit, 100, 2, 7);
        extract_and_disp(counter_limit, 10, 2, 8);
        extract_and_disp(counter_limit, 1, 2, 9);
        Lcd_Out(2, 10, "       ");
        if (keypad_btn == up_btn) {
          counter_limit += 10;
          if (counter_limit > 150) {counter_limit = 150;}
        }
        if (keypad_btn == down_btn) {
          counter_limit -= 10;
          if (counter_limit < 10) {counter_limit = 10;}
        }
        break;
      }
      keypad_btn = scan_keypad();          // Scan the keypad
      if (keypad_btn == menu_btn) {
        menu_flag = menu_inactive;         // Deactivate menu system if menu button pressed
        write_eeprom();                    // Write data to EEPROM
        calc_txon();                       // Calculate TX pulse width and set timer
        calc_txpd();                       // Calculate TX period and set timer
        calc_main_dly();                   // Calculate main sample delay and set timer
        calc_main_smpl();                  // Calculate main sample pulse width and set timer
        calc_efe_dly();                    // Calculate EFE sample delay and set timer
        calc_efe_smpl();                   // Calculate EFE sample pulse width and set timer
```

```c
    calc_disc_dly();                                // Calculate disc sample delay and set timer
    calc_disc_smpl();                               // Calculate disc sample pulse width and set timer
    clear_arrays();                                 // Clear both PI and DISC arrays
    int_state = 0;                                  // Reset interrupt state machine
    PIE1 = 0x01;                                     // Enable TMR0 interrupt
    INTCON = 0xE0;                                    // Enable global, peripheral and TMR1 interrupts
  } else {
    if (keypad_btn == enter_btn) {                  // Navigate menu system
      switch (menu_disp) {
        case 0:                                     // Detect mode
          if (detect_mode == hybrid) {
            menu_disp = 6;                           // Go to hybrid settings
          } else {
            menu_disp = 1;                           // Go to pulse settings
          }
          break;
        case 1:                                     // PI threshold
          menu_disp = 2;
          break;
        case 2:                                     // PI pulse width
          menu_disp = 3;
          break;
        case 3:                                     // PI sample delay
          menu_disp = 4;
          break;
        case 4:                                     // PI sample pulse width
          menu_disp = 5;
          break;
        case 5:                                     // PI running average
          menu_disp = 0;
          break;
        case 6:                                     // Disc threshold
          menu_disp = 7;
          break;
        case 7:                                     // Disc sample delay
          menu_disp = 8;
          break;
        case 8:                                     // Disc sample pulse width
          menu_disp = 9;
          break;
        case 9:
          menu_disp = 10;                            // Disc running average
          break;
        case 10:                                    // Meter zero limit
          menu_disp = 11;
          break;
        case 11:                                    // Accept and reject counter limit
          menu_disp = 0;
          break;
      }
    }
  }
}

void disp_reset() {
  if (detect_mode == pulse) {
    Lcd_Out(1, 1, " PULSE   ");
  } else {
    Lcd_Out(1, 1, " HYBRID  ");
  }
  Lcd_Out(2, 1, "        ");
  Lcd_Chr(2, 8, 7);                                 // Display middle marker (vertical line)
  Lcd_Out(2, 9, "        ");
}

// Main program section
void main() {
  // Initialize ports
  TRISA = TRISA_INIT;
  TRISB = TRISB_INIT;
  TRISC = TRISC_INIT;
```

The Voodoo Project

```
TRISD = TRISD_INIT;
TRISE = TRISE_INIT;

// Turn off MOSFET and sample pulses
EN1 = EN1_off;                                      // Turn off MOSFET
main_pulse = main_off;                              // Turn off main sample pulse
efe_pulse = efe_off;                                // Turn off EFE sample pulse
disc_pulse = disc_off;                              // Turn off disc sample pulse

// Initialize ADCs
ADCON1 = 0x07;                                      // Enable AD0 to AD7 (VDD and VSS as voltage ref)
ADC_Init();

// Initialize LCD
Lcd_Init();
Lcd_Cmd(_LCD_CLEAR);
Lcd_Cmd(_LCD_CURSOR_OFF);

// Load custom characters into CG RAM of LCD
Lcd_Cmd(64);
// Battery >= 12.0V (character 0)
Lcd_Chr_Cp(14);
for (i = 2; i <= 8; i++) {Lcd_Chr_Cp(31);}
// Battery >= 11.7V (character 1)
Lcd_Chr_Cp(14);
Lcd_Chr_Cp(31);
Lcd_Chr_Cp(17);
for (i = 4; i <= 8; i++) {Lcd_Chr_Cp(31);}
// Battery >= 11.4V (character 2)
Lcd_Chr_Cp(14);
Lcd_Chr_Cp(31);
for (i = 3; i <= 4; i++) {Lcd_Chr_Cp(17);}
for (i = 5; i <= 8; i++) {Lcd_Chr_Cp(31);}
// Battery >= 11.1V (character 3)
Lcd_Chr_Cp(14);
Lcd_Chr_Cp(31);
for (i = 3; i <= 5; i++) {Lcd_Chr_Cp(17);}
for (i = 6; i <= 8; i++) {Lcd_Chr_Cp(31);}
// Battery >= 10.8V (character 4)
Lcd_Chr_Cp(14);
Lcd_Chr_Cp(31);
for (i = 3; i <= 6; i++) {Lcd_Chr_Cp(17);}
for (i = 7; i <= 8; i++) {Lcd_Chr_Cp(31);}
// Battery >= 10.5V (character 5)
Lcd_Chr_Cp(14);
Lcd_Chr_Cp(31);
for (i = 3; i <= 7; i++) {Lcd_Chr_Cp(17);}
Lcd_Chr_Cp(31);
// Block symbol for ferrous /non-ferrous display (character 6)
for (i = 1; i <= 8; i++) {Lcd_Chr_Cp(31);}
// Middle character on ferrous / non-ferrous display (character 7)
for (i = 1; i <= 8; i++) {Lcd_Chr_Cp(4);}

// Display splash screen
Lcd_Out(1,1," VOODOO V1.0   ");
Lcd_Out(2,1,"HYBRID DETECTOR ");
delay_ms(2000);
Lcd_Cmd(_LCD_CLEAR);

// Initialize variables
int_state = 0;                                      // Initialize interrupt state machine
main_state = 0;                                     // Initialize main state machine
loop_count = 0;                                     // Reset loop counter
hybrid_cycle = 0;                                   // Hybrid operating cycle (0 = pulse, 1 = disc)
accept = 0;                                         // Clear accept (non-ferrous) counter
accept_blk = 0;                                     // Clear accept block display counter
reject = 0;                                         // Clear reject (ferrous) counter
reject_blk = 0;                                     // Clear reject block display counter
update_disp = 0;                                    // Clear update display flag
meter_zero = 0;                                     // Clear meter zero counter

// All default settings below may be overwritten by EEPROM
```

```
detect_mode = hybrid;                           // Set default detection mode to hybrid
pi_thr = 525;                                   // Set default pi threshold
disc_thr = 511;                                 // Set default disc threshold
txon = 150;                                      // Set default TX pulse width (us)
txpd = 1000;                                      // Set default TX pulse period (us)
main_dly = 27;                                   // Set default main sample delay (us)
main_smpl = 60;                                   // Set default main sample pulse width (us)
efe_dly = 650;                                    // Set default EFE sample delay (us)
pi_num = 64;                                       // Set default number of PI readings to average
disc_dly = 45;                                    // Set default disc sample delay (us)
disc_smpl = 45;                                   // Set default disc sample pulse width (us)
disc_num = 64;                                     // Set default number of disc readings to average
meter_zero_limit = 10;                            // Set meter zero limit
counter_limit = 100;                              // Set accept and reject counter limit

clear_arrays();                                    // Clear both PI and DISC arrays

// Load saved values from EEPROM
if (EEPROM_Read(0) == 0xAA) {                     // Check if EEPROM already initialized
  detect_mode = EEPROM_Read(1);                   // Read detection mode (pulse or hybrid)
  pi_thr = EEPROM_Read(2);                        // Read PI threshold high byte
  pi_thr = pi_thr << 8;
  pi_thr += EEPROM_Read(3);                       // Read PI threshold low byte
  txon = EEPROM_Read(4);                          // Read TX pulse width
  txpd = EEPROM_Read(5);                          // Read TX period high byte
  txpd = txpd << 8;
  txpd += EEPROM_Read(6);                         // Read TX period low byte
  main_dly = EEPROM_Read(7);                      // Read main sample delay
  main_smpl = EEPROM_Read(8);                     // Read main sample pulse width
  efe_dly = EEPROM_Read(9);                       // Read EFE sample delay high byte
  efe_dly = efe_dly << 8;
  efe_dly += EEPROM_Read(10);                     // Read EFE sample delay low byte
  pi_num = EEPROM_Read(11);                       // Read number of PI readings to average
  disc_thr = EEPROM_Read(12);                     // Read disc threshold high byte
  disc_thr = disc_thr << 8;
  disc_thr += EEPROM_Read(13);                    // Read disc threshold low byte
  disc_dly = EEPROM_Read(14);                     // Read disc sample delay
  disc_smpl = EEPROM_Read(15);                    // Read disc sample pulse width
  disc_num = EEPROM_Read(16);                     // Read number of disc readings to average
  meter_zero_limit = EEPROM_Read(17);            // Read meter zero limit
  counter_limit = EEPROM_Read(18);               // Read accept and reject counter limit
} else{
  write_eeprom();                                 // Write settings to EEPROM
}
calc_txon();                                       // Calculate TX pulse width and set timer
calc_txpd();                                       // Calculate TX period and set timer
calc_main_dly();                                   // Calculate main sample delay and set timer
calc_main_smpl();                                  // Calculate main sample pulse width and set timer
calc_efe_dly();                                    // Calculate EFE sample delay and set timer
calc_efe_smpl();                                   // Calculate EFE sample pulse width and set timer
calc_disc_dly();                                   // Calculate disc sample delay and set timer
calc_disc_smpl();                                  // Calculate disc sample pulse width and set timer
disp_reset();                                      // Reset display

// Initialize timers and interrupts
// TMR0 used to generate sample pulses and delays (main and EFE) and TX pulses
// TMR1 used to generate TX period
T0CON = 0x88;                                      // Configure TMR0 as 16-bit
T1CON = 0x85;                                      // Configure TMR1 as 16-bit
TMR0H = txonh;                                      // Load TX pulse width
TMR0L = txonl;
TMR1H = txpdh;                                      // Load TX period
TMR1H = txpdl;
PIR1 = 0x00;                                        // Clear TMR1 interrupt overflow flag
PIE1 = 0x01;                                        // Enable TMR1 overflow interrupt
INTCON = 0xE0;                                       // Enable interrupts

while(1) {
  switch (main_state) {
    case 0:
      pi_target = Adc_Read(pi_adc);               // Acquire PI target reading
      pi_tmp = pi_array[pi_pointer];              // Save current PI array value
```

```
pi_array[pi_pointer] = pi_target;        // Add latest reading to PI array
pi_total = (pi_total + pi_target) - pi_tmp; // Calculate PI array total
pi_target = pi_total / pi_num;           // Average readings
pi_pointer++;                            // Increment PI pointer
if (pi_pointer >= pi_num) {
  pi_pointer = 0;                        // Reset PI array pointer
}
if (detect_mode == hybrid) {
  disc_target = Adc_Read(disc_adc);      // Acquire DISC target reading
  disc_tmp = disc_array[disc_pointer];   // Save current disc array value
  disc_array[disc_pointer] = disc_target; // Add latest reading to disc array
  disc_total = (disc_total + disc_target) - disc_tmp; // Calculate disc array total;
  disc_target = disc_total / disc_num;   // Average readings
  disc_pointer++;                        // Increment disc pointer
  if (disc_pointer >= disc_num) {
    disc_pointer = 0;                    // Reset disc array pointer
  }
}

if (pi_target >= pi_thr) {               // Decide whether to beep or not
  if (detect_mode == hybrid) {           // Check for hybrid mode
    update_disp = 0;                     // Reset update display flag
    if (disc_target >= disc_thr) {       // Non-ferrous target detected
      audio_en = audio_on;               // Beep
      if (reject_blk == 0) {
        accept++;                        // Increment accept counter if no reject blocks
        reject = 0;                      // Clear reject counter
        if (accept > counter_limit) {    // Check if accept counter has reached limit
          accept = 0;                    // Reset accept counter to zero
          accept_blk++;                  // Increment number of accept blocks
          if (accept_blk <= blk_limit) {
            update_disp = 1;             // Set update display flag
          } else {
            accept_blk = blk_limit;      // Limit number of accept blocks
          }
        }
      } else {                           // There must be some reject blocks displayed
        if (reject != 0) {
          reject--;                      // Decrement reject counter if not at zero
        } else {                         // Reject counter must have reached zero
          reject = counter_limit;        // Set reject counter to limit
          if (reject_blk != 0) {
            reject_blk--;                // Decrement number of reject blocks if not zero
            update_disp = 1;             // Set update display flag
          }
        }
      }
    } else {                             // Ferrous target detected
      audio_en = audio_off;              // Do not beep
      if (accept_blk == 0) {
        reject++;                        // Increment reject counter if no accept blocks
        accept = 0;                      // Clear accept counter
        if (reject > counter_limit) {    // Check if reject counter has reached limit
          reject = 0;                    // Reset reject counter to zero
          reject_blk++;                  // Increment number of reject blocks
          if (reject_blk <= blk_limit) {
            update_disp = 1;             // Set update display flag
          } else {
            reject_blk = blk_limit;      // Limit number of reject blocks
          }
        }
      } else {                           // There must be some accept blocks displayed
        if (accept != 0) {
          accept--;                      // Decrement accept counter is not at zero
        } else {                         // Accept counter must have reached zero
          accept = counter_limit;        // Set accept counter to limit
          if (accept_blk != 0) {
            accept_blk--;                // Decrement number of accept blocks if not zero
            update_disp = 1;             // Set update display flag
          }
        }
      }
    }
```

```
      }
    } else {                                          // Mode must be PI
      audio_en = audio_on;                            // Beep
    }
  } else {                                            // No target detected
    audio_en = audio_off;                             // Do not beep
  }
  if (update_disp == 0) {                             // Only zero display when no signal is present
    meter_zero++;                                     // Increment meter zero counter
    if (meter_zero >= meter_zero_limit) {
      if (accept_blk != 0) {                          // Must be accept blocks displayed
        if (accept != 0) {
          accept--;                                   // Decrement accept counter if not already zero
        } else {
          accept_blk--;                               // Decrement number of accept blocks
          accept = counter_limit;                     // Set accept counter to limit
          update_disp = 1;                            // Set update display flag
        }
      } else {
        if (reject_blk != 0) {                        // Otherwise must be reject blocks displayed
          if (reject != 0) {
            reject--;                                 // Decrement reject counter if not already zero
          } else {
            reject_blk--;                             // Decrement number of reject blocks displayed
            reject = counter_limit;                   // Set reject counter to limit
            update_disp = 1;                          // Set update display flag
          }
        }
      }
      meter_zero = 0;                                 // Reset meter zero counter
    }
  }
  if (update_disp == 1) {                             // Check if display needs to be updated
    if (accept_blk != 0) {
      for (i = 1; i <= accept_blk; i++) {
        Lcd_Chr(2, 8 + i, 6);                         // Display required number of non-ferrous blocks
      }
      if (accept_blk != blk_limit) {
        for (i = accept_blk + 1; i <= blk_limit; i++) {
          Lcd_Chr(2, 8 + i, 32);                      // Fill rest of non-ferrous display with spaces
        }
      }
    } else {
      Lcd_Chr(2, 9, 32);                              // Clear first block in non-ferrous display
      if (reject_blk != 0) {
        for (i = 1; i <= reject_blk; i++) {
          Lcd_Chr(2, 8 - i, 6);                       // Display required number of ferrous blocks
        }
        if (reject_blk != blk_limit) {
          for (i = reject_blk + 1; i <= blk_limit; i++) {
            Lcd_Chr(2, 8 - i, 32);                    // Fill rest of ferrous display with spaces
          }
        }
      } else {
        Lcd_Chr(2, 7, 32);                            // Clear first block in ferrous display
      }
    }
    update_disp = 0;                                  // Reset update display flag
  }
  keypad_btn = scan_keypad();
  if (keypad_btn == menu_btn) {                       // Check if menu button pressed
    menu_system();                                    // Enter menu system
    disp_reset();                                     // Reset display after exiting menu system
  }
  loop_count++;                                       // Increment loop count
  main_state = (loop_count == 128)?1:0;               // Read battery voltage after loop count of 128
  break;
case 1:
  battery_measure();                                  // Acquire battery voltage
  battery_display();                                  // Display battery voltage
  battery_symbol();                                   // Display appropriate battery symbol
```

The Voodoo Project

```
            loop_count = 0;                  // Reset loop count
            main_state = 0;                  // Reset main program state machine
            break;
        }
    }
}
```